# Adaptive Filter Algorithms For Complex and Quaternion Data

Thomas K. Paul · Tokunbo Ogunfunmi

# Adaptive Filter Algorithms For Complex and Quaternion Data

Springer

Thomas K. Paul
Department of Electrical & Computer Engineering
Santa Clara University
Santa Clara, CA, USA

Tokunbo Ogunfunmi
Department of Electrical & Computer Engineering
Santa Clara University
Santa Clara, CA, USA

ISBN 978-3-032-23797-2 ISBN 978-3-032-23798-9 (eBook)
https://doi.org/10.1007/978-3-032-23798-9



This Springer imprint is published by the registered company Springer Nature Switzerland AG
The registered company address is: Gewerbestrasse 11, 6330 Cham, Switzerland

# Preface

Adaptive filters may be described as a class of filtering structures having a mechanism for adjusting the free parameters (i.e. coefficients) based on statistical variations in the environment where the filter operates [1]. Adaptive filters have numerous applications, including adaptive equalization, adaptive noise cancellation, beamforming, system identification, etc. Adaptive filters were originally developed for real signals. Later, an extension to the complex domain was published. In recent years, adaptive filters have been developed for trinions, quaternions, octonions and hyper-complex data.

For applications in communications, image processing and biological systems, nonlinear filters are well-suited for modeling the naturally occurring physical characteristics. There is an interest in understanding methods of nonlinear adaptive filtering. Kernel methods are a powerful new approach that can provide considerable complexity reduction for nonlinear filtering, as well as the potential for universal function approximation. The approach is of particular interest for real-time applications.

Quaternions, which are an example of hyper-complex data, provide a convenient representation for performing transformations in 3- or 4-dimensional space. Quaternions are particularly well-suited for robotics applications, specifically image pattern recognition and modeling motion in 3-dimensions. However, the use of quaternions for adaptive filtering has been hampered by the fact that multiplication of quaternions is non-commutative, which complicates gradient operations using quaternion-based cost functions.

In this book, we present linear and nonlinear adaptive algorithms for use with complex and quaternion-valued data. We use kernel adaptive algorithms in cases where nonlinearity is present.

Specifically, we present the complex kernel method: Complex Kernel Affine Projection Algorithm (CKAPA), and complex kernel filtering enhancement using widely linear (WL) estimation (i.e. derived WL-CKLMS, WL-CKAPA). For quaternion-based kernel methods, we present algorithms such as the Quaternion Kernel LMS (Quat-KLMS) algorithm and Quaternion Kernel Maximum Correntropy (Quat-KMC) algorithm. We present convergence analysis of the algorithms and use simulation to verify the theoretical derivation using various applications of the adaptive filters.

The book is organized as follows: Chap. 1 is an introduction, while in Chap. 2, we discuss linear and kernel filter theory, including reproducing kernels and transform spaces applied to adaptive filters. Chapter 3 is about adaptive filters for complex data. Chapter 4 discusses the complex kernel Least Mean Square (CKLMS) adaptive algorithm. Chapter 5 introduces use of quaternions for adaptive filters, while Chap. 6 is an introduction to Information Theoretic Learning. The derivation of the Complex Kernel APA algorithm is in Chap. 7. Chapter 8 presents the development of the Widely Linear CKLMS, CKAPA algorithms. Convergence analysis of the Complex KLMS algorithm is presented in Chap. 9. The derivation of the Quaternion Kernel LMS algorithm is presented in Chap. 10, while Chap. 11 has the derivation of the Quaternion Kernel Maximum Correntropy algorithm.

We hope the material here will help educate readers and highlight some of the recent contributions to the field of adaptive signal processing using complex and quaternion data.

We thank our families for their love and support during the time to prepare and publish this manuscript.

Santa Clara, CA, USA

Tokunbo Ogunfunmi
Thomas K. Paul

# Contents

# Chapter 1
# Introduction

## 1.1 Real, Complex and Quaternion Numbers

Adaptive signal processing algorithms have been developed originally for real numbers, R, and then later for complex numbers, C. Later, the algorithms have been extended to quaternions. This book is about presenting a unified view of the developments of these algorithms for complex and quaternion numbers. In this chapter, we introduce real, complex and quaternion number systems. Other number systems of interest are trinions, octonions and hypercomplex numbers. These topics are not discussed here.

## 1.2 Number Systems

Real numbers have been known for many years. Problems whose solutions require finding square roots of negative numbers led to the idea of complex numbers. Mathematicians later worked on the problem of solving for the roots of an arbitrary polynomial, leading to the Fundamental Theorem of Algebra. The theorem, as may be currently understood, states that every nth order polynomial with real coefficients has exactly $n$ roots in the complex domain, C. For example, the polynomial expression:

$$ay^3 + by^2 + cy + d = 0$$

has exactly three roots in C.

Work in this area by several mathematicians, including Descartes, John Wallis, Leonhard Euler, Abraham De Moivre and Karl Friedrich Gauss, resulted in the development of what is now known as complex numbers.



T. K. Paul and T. Ogunfunmi, *Adaptive Filter Algorithms For Complex and Quaternion Data*, https://doi.org/10.1007/978-3-032-23798-9_1

## 1.3 Quaternion Numbers

Quaternions were introduced by Sir William Rowan Hamilton, who discovered them in 1843 [8]. During this time, he set out to invent an algebra of "triplets" to perform similar operations in three-dimensional space as complex numbers were found to perform in two-dimensional space.

In the process of developing this algebra, Hamilton defined the fundamental equations between i, j, k, as follows:

$$i^2 = j^2 = k^2 = ijk = -1 \tag{1.1}$$

Any quaternion value can be expressed as

$$q = q_a + iq_b + jq_c + kq_d = Sq + Vq \tag{1.2}$$

where $\text{R}(q) = Sq = q_a$ is the scalar (real) part, and the vector (or pure quaternion) part is $\text{J}(q) = Vq = iq_b + jq_c + kq_d$, and where $q_a$, $q_b$, $q_c$ and $q_d$ are all real-valued coefficients.

In particular, the following properties can be shown based on (1.1):

$$ij = -ji = k \quad jk = -kj = i \quad ki = -ik = j \tag{1.3}$$

In the next section, we describe how quaternions defined in this fashion may be used for modeling rotations in three-dimensional space.

Trinions, Octonions and Hypercomplex numbers [1–10] are not discussed in this book. However, more details of quaternions are presented in Chap. 5.

## 1.4 Adaptive Filters

The use of complex data for adaptive filters is known to provide benefits for many applications (communication systems modeling, image processing, etc.) due to its ability to represent amplitude and phase information. Meanwhile, quaternion data provides benefit in performing transformations in 3 or 4-dimensional space, and may be applied to pattern recognition, motion tracking applications, etc. The research areas of complex and quaternion-valued data with kernel adaptive methods have gained much interest, some of which are presented in this book.

## 1.5 Purpose of This Work

The purpose of this book is to aid the understanding of, as well as develop algorithms for the use of complex and quaternion data with kernel methods. Specific objectives were: (1) elucidate and enhance the transform mappings used for nonlinear system learning using complex or quaternion data. This involves considering the choice of kernel parameters based on convergence and MSE, as well as the design of alternative kernel functions. (2) present applications used to boost the performance of kernel algorithms. Example areas of research include widely linear estimation, information theoretic learning and analyzing the performance and convergence behavior of kernel adaptive methods using complex data.

## 1.6 Organization of the Book

This book is organized as follows. The following five chapters (Chaps. 2–6) describe background material that formed the foundation of the algorithms presented here. The areas covered in these chapters are.

- Linear and kernel adaptive filtering (Chap. 2)
- Use of complex data for adaptive filters (Chap. 3)
- Description of the Complex Kernel LMS algorithm (Chap. 4)
- Quaternions for adaptive filters (Chap. 5)
- Information Theoretic Learning (Chap. 6)

The remaining chapters (Chaps. 7–11) describe the analysis and algorithms developed and presented here. These are

- Derivation of the Complex Kernel APA (Chap. 7)
- Development of Widely Linear CKLMS, CKAPA (Chap. 8)
- Convergence analysis of the Complex KLMS (Chap. 9)
- Derivation of the Quaternion Kernel LMS (Chap. 10)
- Derivation of Quaternion Kernel Max. Correntropy algorithm (Chap. 11)

## References


1. J. Huerta, "*Introducing the Quaternions*". http://math.ucr.edu/~huerta/introquaternions.pdf
2. M.J. Crowe, "*A history of vector analysis*". https://www.math.ucdavis.edu/~temple/MAT21D/SUPPLEMENTARY-ARTICLES/Crowe_History-of-Vectors.pdf
3. M.J. Crowe, *A history of vector analysis: The evolution of the idea of a vectorial system, published by University of Notre Dame Press* (1967)
4. A. Inhausen, "*Mathematical fragments from Lahun: Introduction*" (University College London, 2002). https://www.ucl.ac.uk/museums-static/digitalegypt//lahun/mathintro.html

5. L. Vicci, "*Quaternions and rotations in 3-space: The algebra and its geometric interpretation*". http://www.cs.unc.edu/techreports/01-014.pdf
6. M. Bahri, Mulkiah, N. Erawati, R. Ashino, "Real matrix representation for trinion numbers and its properties". Inf. Int. Interdiscip. J., **19**(6(A)), 1657–1666 (2016)
7. J.T. Graves, On a connection between the general theory of normal couples and the theory of complete quadratic functions of two variables. Phil. Mag. **26**, 315–320 (1845)
8. W.R. Hamilton, Note, by Sir W. R. Hamilton, respecting the researches of John T. Graves, Esq. Transactions of the Royal Irish Academy **21**, 338–341 (1848)
9. R.A. Wilson, "*Octonions*". http://www.maths.qmul.ac.uk/~raw/talks_files/octonions.pdf
10. R.A. Wilson, *WWW-O The weird and wonderful world of octonions*. http://www.maths.qmul.ac.uk/~raw/talks_files/octonions.pdf

# Chapter 2
# Linear and Kernel Filter Theory

## 2.1 Linear Adaptive Filtering

Linear adaptive filters may be described as a class of filtering structures having a mechanism for adjusting the free parameters (i.e. coefficients) based on statistical variations in the environment where the filter operates [1]. Adaptive filters have numerous applications, including adaptive equalization, adaptive noise cancellation, beamforming, system identification, etc. Figure 2.1 illustrates the structure for a linear adaptive filter.

Adaptive filters typically rely on error-correction learning for adaptation. This is visible from Fig. 2.1, where error $e(n)$ is used by the adaptive mechanism for adjusting the weights $\mathbf{w}(n)$ each iteration $n$. The weights $\mathbf{w}(n)$ usually define a transversal filter on input $\mathbf{u}(n)$.

The adaptive update for the weights can be expressed as

$$\mathbf{w}(n) = \mathbf{w}(n-1) + \Delta\mathbf{w}(n) \tag{2.1}$$

where $\Delta\mathbf{w}(n)$ is the weight increment performed at each iteration.

Some algorithms for updating the weights, which are considered within this work, are described in the following subsections.

### 2.1.1 *Wiener Filtering*

It is well known that a good choice for the filter coefficients $\mathbf{w}$ is one that minimizes the mean square error, or MSE [1–5]. The weights satisfying this criterion may be found as follows. For the MSE, we write

$$E[e^2(n)] = E[(d(n) - \mathbf{w}^T\mathbf{u}(n))^2]$$



T. K. Paul and T. Ogunfunmi, *Adaptive Filter Algorithms For Complex and Quaternion Data*, https://doi.org/10.1007/978-3-032-23798-9_2

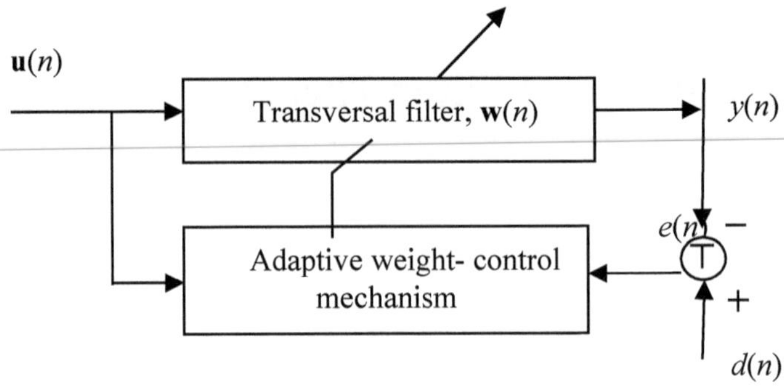


**Fig. 2.1** Example linear adaptive filter structure

$$= E[d^2(n)] - E[d(n)\mathbf{w}^T\mathbf{u}(n)] - E[\mathbf{u}^T(n)\mathbf{w}\,d(n)] + E[\mathbf{w}^T\mathbf{u}(n)\mathbf{u}^T(n)\mathbf{w}] \tag{2.2}$$

where $E[.]$ is the expectation operation. This may be expressed as

$$\begin{aligned} E[e^2(n)] &= E[d^2(n)] - \mathbf{w}^T E[d(n)\mathbf{u}(n)] - E[\mathbf{u}^T(n)d(n)]\mathbf{w} + \mathbf{w}^T E[\mathbf{u}(n)\mathbf{u}^T(n)]\mathbf{w} \\ &= \sigma_d^2 - \mathbf{w}^T\mathbf{p} - \mathbf{p}^T\mathbf{w} + \mathbf{w}^T\mathbf{R}\mathbf{w} \end{aligned} \tag{2.3}$$

where $\mathbf{R} = E[\mathbf{u}(n)\mathbf{u}^T(n)]$ is the autocorrelation matrix of $\mathbf{u}(n)$, and $\mathbf{p} = E[d(n)\mathbf{u}(n)]$ is the cross-correlation vector between $d(n)$ and $\mathbf{u}(n)$.

For minimizing (2.3), we may find the derivative w.r.t. $\mathbf{w}$ and set to zero, yielding

$$\begin{aligned} \frac{\partial}{\partial \mathbf{w}}\{\sigma_d^2 - \mathbf{w}^T\mathbf{p} - \mathbf{p}^T\mathbf{w} + \mathbf{w}^T\mathbf{R}\mathbf{w}\} &= 0 \\ -2\mathbf{p} + 2\mathbf{R}\mathbf{w} &= 0 \\ \mathbf{w} = \mathbf{R}^{-1}\mathbf{p} & \end{aligned} \tag{2.4}$$

This solution for the weights is referred to as the Weiner solution, and is referred to often in this section. The Wiener solution does not describe an adaptive update, it specifies the weights that statistically minimize the mean, or average squared error. Approaches for adaptively updating the filter weights are described in the following subsections.

### *2.1.2 Gradient-Based Descent and Newton's Method*

Here, we describe the gradient-based approaches used for determining filter weight updates. Steepest descent and Newton's method are described. These approaches can be seen to form the foundation for the LMS and APA weight updates derived in later subsections.

The method of steepest descent is a gradient-based iterative descent formed in the following way. Starting from an initial guess $\mathbf{w}(0)$ for the weights, an iterative weight update is computed with the goal of reducing the cost function, $J(\mathbf{w})$. For MSE

$$J(\mathbf{w}) = E[e^2(n)]$$

The approach used is to iteratively update $\mathbf{w}(n)$ in the opposite direction of the gradient $\nabla_{\mathbf{w}} J(\mathbf{w})$, the direction leading to steepest descent [2]. Writing $\mathbf{g} = \nabla_{\mathbf{w}} J(\mathbf{w})$, the weight update is

$$\mathbf{w}(n) = \mathbf{w}(n-1) - \mu\, \mathbf{g}(n-1) \tag{2.5}$$

where $\mu$ is a positive constant known as the step-size, and is used for controlling the speed and stability of convergence, as well as misadjustment.

Using the gradient from (2.4), we have

$$\begin{aligned} \mathbf{w}(n) &= \mathbf{w}(n-1) + \mu[2\mathbf{p} - 2\mathbf{R}\mathbf{w}(n-1)] \\ &= \mathbf{w}(n-1) + \mu[\mathbf{p} - \mathbf{R}\mathbf{w}(n-1)] \end{aligned} \tag{2.6}$$

where the scaling factor 2 is included in step-size $\mu$.

The steepest descent is simple in design, however it may require many iterations for convergence. An alternative is Newton's method, which is based on the second-order Taylor series expansion of the cost function $J(\mathbf{w})$ w.r.t. $\mathbf{w}(n)$, that is [2, 3]:

$$J(\mathbf{w}(n) + \Delta\mathbf{w}(n)) \;=\; J(\mathbf{w}(n)) \;+\; \Delta\mathbf{w}(n)^T \mathbf{g}(n) \;+\; \tfrac{1}{2}\Delta\mathbf{w}(n)^T \mathbf{H}(n) \Delta\mathbf{w}(n) \tag{2.7}$$

where $\Delta\mathbf{w}(n)$ is the weight increment between iterations $n$ and $n+1$, and

$$\mathbf{g}(n) = \left.\frac{\partial J(\mathbf{w})}{\partial \mathbf{w}}\right|_{\mathbf{w}=\mathbf{w}(n)}, \qquad \mathbf{H}(n) = \left.\frac{\partial J^2(\mathbf{w})}{\partial^2 \mathbf{w}}\right|_{\mathbf{w}=\mathbf{w}(n)} \tag{2.8}$$

where vector $\mathbf{g}(n)$ is the gradient at $\mathbf{w}(n)$, and matrix $\mathbf{H}(n)$ the Hessian at $\mathbf{w}(n)$.

Finding the gradient of (2.7) w.r.t. $\Delta\mathbf{w}(n)$, and setting it to zero yields

$$\begin{aligned} \mathbf{g}(n) + \mathbf{H}(n)\Delta\mathbf{w}(n) &= 0 \\ \Delta\mathbf{w}(n) &= -\mathbf{H}^{-1}(n)\,\mathbf{g}(n) \end{aligned} \tag{2.9}$$

Based on (2.3) and (2.8), we also know $\mathbf{H}(n) = 2\mathbf{R}$. The weight update following Newton's method is

$$\begin{aligned}\mathbf{w}(n) &= \mathbf{w}(n-1) + \mu\,\mathbf{H}^{-1}(n-1)\mathbf{g}(n-1)\\ &= \mathbf{w}(n-1) + \mu\,(2\mathbf{R})^{-1}(2\mathbf{p} - 2\mathbf{R}\mathbf{w}(n-1))\\ &= \mathbf{w}(n-1) + \mu\mathbf{R}^{-1}(\mathbf{p} - \mathbf{R}\mathbf{w}(n-1))\end{aligned} \tag{2.10}$$

Note that when $\mu = 1$, the Wiener solution $\mathbf{R}^{-1}\mathbf{p}$ may be reached in a single iteration with Newton's method. However, the added computation for finding the inverse of matrix $\mathbf{R}$ may make the use of Newton's method undesirable compared to steepest descent.

The steepest descent and Newton's method approaches require knowledge of autocorrelation $\mathbf{R}$ and cross-correlation $\mathbf{p}$, which are difficult to know in practice. The LMS and APA are based on more readily available information, and are described next.

### 2.1.3 *Least Mean Square Algorithm*

The Least Mean Square (or LMS) algorithm is the simplest as well as the most widely used adaptive filter algorithm [1–4]. The goal of the LMS algorithm is minimizing the following instantaneous squared error:

$$J(n) = e^2(n) \tag{2.11}$$

Equation (2.11) is referred to as the algorithm cost function. We note that the well-known Weiner solution for the optimal weights is based on minimizing $E[e^2(n)]$. The cost function in Eq. (2.11) is considered minimizing the instantaneous squared error, as opposed to the ensemble-averaged square error.

For minimizing the cost function (2.11), the method of steepest descent is used

$$\mathbf{w}(n) = \mathbf{w}(n-1) - \mu\nabla_{\mathbf{w}}J(n) \tag{2.12}$$

where $\nabla_{\mathbf{w}}$ denotes gradient with respect to $\mathbf{w}(n)$, and $\mu$ is the step-size.

Considering Fig. 2.1, the error can be expressed as

$$e(n) = d(n) - \mathbf{w}^T(n-1)\mathbf{u}(n) \tag{2.13}$$

and the gradient may be written as

$$\begin{aligned}\nabla_{\mathbf{w}}J(n) &= \nabla_{\mathbf{w}}\left[e^2(n)\right]\\ &= \nabla_{\mathbf{w}}\left[\left(d(n) - \mathbf{w}^T(n-1)\mathbf{u}(n)\right)^2\right]\\ &= -2\mathbf{u}(n)\left(d(n) - \mathbf{w}^T(n-1)\mathbf{u}(n)\right)\\ &= -2\mathbf{u}(n)e(n)\end{aligned} \tag{2.14}$$

Therefore, the weight update for LMS has the form

$$\mathbf{w}(n) = \mathbf{w}(n-1) + \mu\, \mathbf{u}(n)e(n) \tag{2.15}$$

where the factor of 2 from the gradient is included in step-size $\mu$.

The LMS algorithm can be seen to have a simple design complexity, and is therefore well-suited for practical implementations. The algorithm has also been shown to provide robust adaptive filter performance compared to other methods.

However, the convergence of the LMS algorithm can often be slow, particularly with colored signals. The Affine Projection Algorithm was developed in order to improve this deficiency.

### 2.1.4 *Affine Projection Algorithm*

The Affine Projection Algorithm (APA) was developed by Ozeki and Umeda [6], and is studied in many references, including [7, 8]. The goal of the algorithm is improving upon the convergence of LMS, particularly for colored noise input, while maintaining comparable algorithm complexity.

The APA algorithm improves convergence by using the previous $K$ input vectors, as opposed to only the current vector used by LMS. The cost function of the algorithm is based on the idea of "minimum disturbance" [2–4], which may be stated as follows.

> Find the new weights $\mathbf{w}(n)$ with minimum distance from the previous weights $\mathbf{w}(n\text{-}1)$, subject to some constraint on the output of the filter using these newly chosen weights $\mathbf{w}(n)$.

The constraint used for the APA is that the new weights are chosen such that the past $K$ errors from observations are forced to zero.

The cost function for the APA can be expressed as

$$\begin{aligned} &Minimize: \ \frac{1}{2}\|\mathbf{w}(n) - \mathbf{w}(n-1)\|^2 \\ &Subject\ to: \ \mathbf{d}_{ap}(n) - \mathbf{X}_{ap}^T(n)\mathbf{w}(n) = 0 \end{aligned} \tag{2.16}$$

where

$$\mathbf{X}_{ap}(n) = [\mathbf{x}(n-K+1)\ \ \mathbf{x}(n-K+2)\ \ ...\ \ \mathbf{x}(n)]$$

$$\mathbf{d}_{ap}(n) = [d(n-K+1)\ \ d(n-K+2)\ \ ...\ \ d(n)]$$

are the past $K$ inputs and past $K$ observations, respectively.

Equation (2.16) is a constrained cost function. In order to convert (2.16) to an unconstrained form, the method of Lagrange multipliers is used. This yields

$$\text{Minimize}: \quad \frac{1}{2}\|\mathbf{w}(n) - \mathbf{w}(n-1)\|^2 + \boldsymbol{\lambda}_{ap}^T \left[\mathbf{d}_{ap}(n) - \mathbf{X}_{ap}^T(n)\mathbf{w}(n)\right] \tag{2.17}$$

where $\boldsymbol{\lambda}_{ap}$ is a $K \times 1$ vector of Lagrange multipliers.

Finding the gradient of (2.17) and solving for the weight update yields [2, 3]

$$\mathbf{w}(n) = \mathbf{w}(n-1) + \mu\, \mathbf{X}_{ap}(n)\left(\mathbf{X}_{ap}^T(n)\mathbf{X}_{ap}(n)\right)^{-1} \mathbf{e}_{ap}(n) \tag{2.18}$$

forming the update for the APA.

### 2.1.5 *Alternative Derivation of APA*

For an alternative derivation of APA, we consider the approach described in [1]. The goal for both the APA and LMS algorithms is based on finding weights that minimize the mean-square error. For the complex input case, we can write

$$\min_{\mathbf{w}} \; E\left|d(n) - \mathbf{w}^H(n)\mathbf{x}(n)\right|^2 \tag{2.19}$$

Based on Sect. 2.1.1, the Weiner solution is given by $\mathbf{w}^o = \mathbf{R_x}^{-1}\mathbf{r}_{\mathrm{dx}}$, where $\mathbf{R_x} = E[\mathbf{x}(n)\mathbf{x}^H(n)]$ is the covariance matrix, and $\mathbf{r}_{\mathrm{dx}} = E[d^*(n)\mathbf{x}(n)]$ is the cross-covariance vector.

The APA and LMS algorithms can be derived based on the gradient descent method, which iteratively estimates $\mathbf{w}^o$. This method can be expressed with equation

$$\mathbf{w}(0) = \textit{initial guess}, \quad \mathbf{w}(n) = \mathbf{w}(n-1) + \mu[\mathbf{r}_{d\mathbf{x}} - \mathbf{R_x}\mathbf{w}(n-1)] \tag{2.20}$$

For improved convergence, the regularized Newton's recursion form can be used, i.e.

$$\mathbf{w}(0) = \textit{initial guess}, \quad \mathbf{w}(n) = \mathbf{w}(n-1) + \mu(\mathbf{R_x} + \varepsilon\mathbf{I})^{-1}[\mathbf{r}_{d\mathbf{x}} - \mathbf{R_x}\mathbf{w}(n-1)] \tag{2.21}$$

where $\varepsilon$ is the regularization constant.

The LMS and APA algorithms may be derived from (2.20) and (2.21) based on the estimates chosen for $\mathbf{R_x}$ and $\mathbf{r}_{\mathrm{dx}}$. For LMS, the estimates $\hat{\mathbf{R}}_\mathbf{x}$ and $\hat{\mathbf{r}}_{d\mathbf{x}}$ are found using only the current input $\mathbf{x}(n)$ and observation $d(n)$.

This results in the following estimates:

$$\hat{\mathbf{R}}_\mathbf{x} = \mathbf{x}(n)\mathbf{x}^H(n), \qquad \hat{\mathbf{r}}_{d\mathbf{x}} = d^*(n)\mathbf{x}(n) \tag{2.22}$$

For APA, the $K$ most recent inputs, observations are used, i.e.

$$\begin{aligned}\mathbf{X}(n) &= [\mathbf{x}(n-K+1)\ \ \mathbf{x}(n-K+2)\ \ldots\ \mathbf{x}(n)]\\ \mathbf{d}(n) &= [d(n-K+1)\ \ d(n-K+2)\ \ldots\ d(n)]^T\end{aligned} \tag{2.23}$$

The estimates for $\mathbf{R}_\mathbf{x}$ and $\mathbf{r}_{\mathrm{d}\mathbf{x}}$ are

$$\begin{aligned}\hat{\mathbf{R}}_\mathbf{x} &= (1/K)\cdot\mathbf{X}(n)\mathbf{X}^H(n)\\ \hat{\mathbf{r}}_{d\mathbf{x}} &= (1/K)\cdot\mathbf{X}(n)\mathbf{d}^*(n)\end{aligned} \tag{2.24}$$

Using (2.24) in (2.20) yields

$$\begin{aligned}\mathbf{w}(n) &= \mathbf{w}(n-1) + \mu[(1/K)\mathbf{X}(n)\mathbf{d}^*(n) - (1/K)\mathbf{X}(n)\mathbf{X}^H(n)\mathbf{w}(n-1)]\\ &= \mathbf{w}(n-1) + \mu\mathbf{X}(n)[\mathbf{d}^*(n) - \mathbf{X}^H(n)\mathbf{w}(n-1)]\end{aligned} \tag{2.25}$$

where $\mu$ includes the scaling $(1/K)$ in the last step of (2.25).

And, using (2.24) in (2.21):

$$\mathbf{w}(n) = \mathbf{w}(n-1) + \mu[\mathbf{X}(n)\mathbf{X}^H(n) + \varepsilon\mathbf{I}]^{-1}\mathbf{X}(n)[\mathbf{d}^*(n) - \mathbf{X}^H(n)\mathbf{w}(n-1)] \tag{2.26}$$

where both $\mu$ and $\varepsilon$ are appropriately scaled.

Considering the matrix inversion lemma, we can write

$$[\mathbf{X}(n)\mathbf{X}^H(n) + \varepsilon\mathbf{I}]^{-1}\mathbf{X}(n) = \mathbf{X}(n)[\mathbf{X}^H(n)\mathbf{X}(n) + \varepsilon\mathbf{I}]^{-1}$$

Equation (2.26) can equivalently be written as

$$\mathbf{w}(n) = \mathbf{w}(n-1) + \mu\mathbf{X}(n)[\mathbf{X}^H(n)\mathbf{X}(n) + \varepsilon\mathbf{I}]^{-1}[\mathbf{d}^*(n) - \mathbf{X}^H(n)\mathbf{w}(n-1)] \tag{2.27}$$

Equations (2.25) and (2.27) are different APA forms referred to as APA-1 and APA-2, respectively in [1].

Other APA forms may be obtained by considering a regularized version of the cost function (2.19). The cost function becomes

$$\min_{\mathbf{w}}\ E\left|d(n) - \mathbf{w}^H(n)\mathbf{x}(n)\right|^2 + \lambda\|\mathbf{w}(n)\|^2 \tag{2.28}$$

Note that $\lambda$ is a different regularization constant than $\varepsilon$ in the regularized Newton's recursion of (2.21).

The gradient method, based on (2.28), is

$$\begin{aligned}\mathbf{w}(n) &= \mathbf{w}(n-1) + \eta[\mathbf{r}_{d\mathbf{x}} - (\lambda\mathbf{I} + \mathbf{R}_\mathbf{x})\mathbf{w}(n-1)]\\ &= (1-\eta\lambda)\mathbf{w}(n-1) + \eta[\mathbf{r}_{d\mathbf{x}} - \mathbf{R}_\mathbf{x}\mathbf{w}(n-1)]\end{aligned} \tag{2.29}$$

and Newton's recursion (using $\varepsilon = 0$).

$$\begin{aligned}\mathbf{w}(n) &= \mathbf{w}(n-1) + \eta(\lambda\mathbf{I} + \mathbf{R}_{\mathbf{x}})^{-1}[\mathbf{r}_{d\mathbf{x}} - (\lambda\mathbf{I} + \mathbf{R}_{\mathbf{x}})\mathbf{w}(n-1)] \\ &= (1-\eta)\mathbf{w}(n-1) + \eta(\lambda\mathbf{I} + \mathbf{R}_{\mathbf{x}})^{-1}\mathbf{r}_{d\mathbf{x}}\end{aligned} \tag{2.30}$$

Substituting (2.24) into (2.29), (2.30) yields

$$\mathbf{w}(n) = (1-\eta\lambda)\mathbf{w}(n-1) + \eta\mathbf{X}(n)[\mathbf{d}^*(n) - \mathbf{X}^H(n)\mathbf{w}(n-1)] \tag{2.31}$$

$$\mathbf{w}(n) = (1-\eta)\mathbf{w}(n-1) + \eta\mathbf{X}(n)[\lambda\mathbf{I} + \mathbf{X}^H(n)\mathbf{X}(n)]^{-1}\mathbf{d}^*(n) \tag{2.32}$$

which will be referred to as APA-3 and APA-4. The derivation of (2.32) also considered the matrix inversion lemma.

Algorithms APA-1 and APA-2 differ in that APA-2 provides improved convergence behavior considering the added Newton's recursion for the gradient descent. This convergence benefit comes at the cost of computational complexity, based on the matrix inversion.

The APA-3, APA-4 are the corresponding algorithms using a regularized cost function. This can be seen to prevent overfitting of the solution (where the solution is chosen based on an exact fit with data) by constraining the weight vector norm. However, this approach may provide a reduced convergence rate versus APA-1, APA-2. These algorithms are the four possible APA variants considered here.

## 2.2 Reproducing Kernels and Transform Spaces

Before describing kernel adaptive filtering, we will initially describe kernel functions (also known as reproducing kernels), as well as their associated transform spaces, referred to as reproducing kernel Hilbert spaces, or RKHS. These functions and their transform spaces, which are nonlinear spaces, allow for efficiently extending linear adaptive filtering techniques for nonlinear filtering. The types of nonlinearities that may be handled are dependent on the design of the kernel functions and transform spaces used. Many references cover these areas, including [1, 9–18]. This section provides background for this.

For understanding these transform spaces, consider an inner product space formed from an orthonormal basis $\{\mathbf{x}_k\}_{k=1}^{\infty}$. This is known as a pre-Hilbert space [1]. Define $\mathcal{H}$ as the most inclusive space of vectors where $\{\mathbf{x}_k\}_{k=1}^{\infty}$ is a basis. The vectors of $\mathcal{H}$ not present in the original pre-Hilbert space can be represented using

$$\mathbf{x} = \sum_{k=1}^{\infty} a_k \mathbf{x}_k \tag{2.33}$$

and are considered spanned by the basis $\{\mathbf{x}_k\}_{k=1}^{\infty}$.

If we define two vectors (spanned by only $n$ and $m$ basis vectors, respectively)

$$y_n = \sum_{k=1}^{n} a_k \mathbf{x}_k, \qquad y_m = \sum_{k=1}^{m} a_k \mathbf{x}_k \tag{2.34}$$

where $n > m$, we can write the squared Euclidean distance of $y_n$ and $y_m$ as

$$\begin{aligned} \|\mathbf{y}_n - \mathbf{y}_m\| &= \left\| \sum_{k=1}^{n} a_k \mathbf{x}_k - \sum_{k=1}^{m} a_k \mathbf{x}_k \right\|^2 = \left\| \sum_{k=m+1}^{n} a_k \mathbf{x}_k \right\|^2 \\ &= \sum_{k=m+1}^{n} a_k^2 \end{aligned} \tag{2.35}$$

since the $\mathbf{x}_k$ vectors are orthogonal.

For the definition (2.33) to be meaningful, we need.

1. $\sum_{k=m+1}^{n} a_k^2 \to 0$, as both $n, m \to \infty$.
2. $\sum_{k=1}^{m} a_k^2 < \infty$.

A sequence of vectors $\{\mathbf{y}_k\}_{k=1}^{\infty}$ defined this way is a Cauchy sequence. A vector $\mathbf{x}$ may be represented with basis $\{\mathbf{x}_k\}_{k=1}^{\infty}$ if and only if $\mathbf{x}$ is a linear combination of the basis vectors, and coefficients $\{a_k\}_{k=1}^{\infty}$ are square summable.

Based on this, we can make the following statement [1].

> An inner product space $\mathcal{H}$ is complete if every Cauchy sequence of vectors taken from the space $\mathcal{H}$ converges to a limit in $\mathcal{H}$; a complete inner product space is called a Hilbert space.

Hilbert spaces are beneficial for learning, since the use of calculus gradient operations can be seen to be applicable to these spaces.

To describe kernel functions used for transforming inputs to Hilbert spaces, consider the following. It has been shown that a kernel function satisfies the following condition, referred to as the Mercer condition [1, 9, 10]:

$$\kappa(\mathbf{a}, \mathbf{b}) = \varphi(\mathbf{a})^T \varphi(\mathbf{b}) \tag{2.36}$$

where $\varphi(\mathbf{x})$ is a mapping function transforming input $\mathbf{x}$ to a Hilbert space (denote this by $\mathcal{H}$). The kernel function $\kappa(\,.\,,\;.\,)$ is also known as a reproducing kernel, and its associated Hilbert space $\mathcal{H}$ (where $\varphi(.)$ is the mapping) referred to as a reproducing kernel Hilbert space, or RKHS [1].

The kernel function $\kappa(\mathbf{a}, \mathbf{b})$ can also be seen to define an inner product for the Hilbert space, $\mathcal{H}$. This can be expressed as $\kappa(\mathbf{a}, \mathbf{b}) = \, < \varphi(\mathbf{b}), \varphi(\mathbf{a}) >_{\mathcal{H}}$. For real-valued kernels, the function is symmetric, or: $< \varphi(\mathbf{b}), \varphi(\mathbf{a}) >_{\mathcal{H}} = \, < \varphi(\mathbf{a}), \varphi(\mathbf{b}) >_{\mathcal{H}}$.

This section only provides a basic description of kernel functions and their associated RKHSs for understanding the chapters to follow. More complete definitions of kernel functions, RKHS are shown in Chap. 4, in the context of complex kernels.

To illustrate, we consider the following kernel functions used in practice:

$$\text{Gaussian kernel}: \quad \kappa(\mathbf{a}, \mathbf{b}) = \exp(\sigma \|\mathbf{a} - \mathbf{b}\|^2) \tag{2.37}$$

$$\text{Polynomial kernel}: \kappa(\mathbf{a}, \mathbf{b}) = (1 + \mathbf{a}^T \mathbf{b})^q$$

where $\sigma$, $q$ are referred to as kernel parameters, and specify the RKHS. In particular, parameter $q$ for the polynomial kernel controls the degree of the polynomial to be learned.

The associated RKHS for the polynomial kernel may be described as follows. For a description of the Gaussian kernel RKHS, see Appendix A. We begin by specifying input vectors **a**, **b** with size $m = 2$, and the parameter $q = 2$. The result can be extended for larger $m$, $q$. In this case, we have

$$\begin{aligned}\kappa(\mathbf{a}, \mathbf{b}) &= (1 + \mathbf{a}^T \mathbf{b})^2 \\ &= (1 + a_1 b_1 + a_2 b_2)^2 \\ &= 1 + 2a_1 b_1 + 2a_2 b_2 + 2a_1 a_2 b_1 b_2 + a_1^2 b_1^2 + a_2^2 b_2^2 \\ &= \varphi(\mathbf{a})^T \varphi(\mathbf{b})\end{aligned} \tag{2.38}$$

where $\varphi(\mathbf{x}) = [1 \ \ \sqrt{2}x_1 \ \ \sqrt{2}x_2 \ \ \sqrt{2}x_1x_2 \ \ x_1^2 \ \ x_2^2]$. Therefore, $\kappa(\mathbf{a}, \mathbf{b})$ is a Mercer kernel, and $\varphi(.)$ is the transform map for the polynomial kernel when $q = 2$ and **a**, **b** are length $m = 2$.

From the resulting transform space, visible from $\varphi(\mathbf{x})$, we see that the input was mapped from the original space of dimension size 2, to a high-dimensional space of dimension size 6. From the dimensions of $\varphi(\mathbf{x})$, we also see that linear coefficients may be applied in this space to learn any nonlinearity of polynomial order 2.

Specifically, defining vector $\mathbf{w} = [w_0 \ldots w_5]$ (dimension size 6), we can write

$$\mathbf{w}^T \varphi(\mathbf{x}) = w_0 + w_1[\sqrt{2}x_1] + w_2[\sqrt{2}x_2] + w_3[\sqrt{2}x_1x_2] + w_4[x_1^2] + w_5[x_2^2] \tag{2.39}$$

Therefore, linear weights may be applied to data in the RKHS for learning nonlinear functions. The use of kernel functions can be seen to provide computational efficiency. To understand this, consider that the dimension size of $\varphi(\mathbf{x})$ is exponential in $q$ for the polynomial kernel, and infinite-dimensional for the Gaussian kernel. However, kernel function evaluations are considerably lower in complexity. Therefore, algorithms can be formed in terms of inner products, which is demonstrated later in the chapter.

## 2.3 Kernel Adaptive Filters

Figure 2.2 shows the block model of a kernel adaptive filter for system identification.

The goal of the kernel adaptive filter can be stated as finding filter $\mathbf{w}(n)$ to estimate the desired response $d(n)$, where $d(n) = f(\mathbf{u}(n)) + v(n)$, with $f(.)$ being a nonlinear function, and where $\mathbf{u}(n)$ is the system input, $v(n)$ is additive noise. The application here may be seen to be similar to the system identification task for linear filters, aside from the following:

(1) The channel $f(.)$ to be learned is nonlinear
(2) The input $\mathbf{u}(n)$ is mapped to an RKHS using $\varphi(.)$.

The mapping of the input to an RKHS allows for the learning of a nonlinear channel using the linear filter $\mathbf{w}(n)$ (see the discussion from the previous section). Considering $\mathbf{w}$(n) is a vector in a Hilbert space, specifically an RKHS, gradient methods may be used for finding the optimal weights.

The estimate $\hat{d}(n)$ may be expressed as

$$\hat{d}(n) = < \varphi(\mathbf{u}(n)), \mathbf{w}(n) >_{\mathcal{H}} \tag{2.40}$$

Many approaches may be used for updating the weights $\mathbf{w}(n)$ for the kernel adaptive filter. An example approach, the Kernel Least Mean Square (KLMS) algorithm [1, 18] is described next.

### 2.3.1 Kernel Least Mean Square Algorithm

The Kernel Least Mean Square (KLMS) algorithm was introduced in [18] and is formed based on using the LMS approach for updating the linear adaptive filter of Fig. 2.2. The KLMS is well-suited compared to LMS for learning when the mapping between $d(n)$, $\mathbf{u}(n)$ is highly nonlinear. The method is described here.

The steepest descent method is used (see Sects. 2.1.2 and 2.1.3), and the weight update can be expressed

$$\mathbf{w}(n) = \mathbf{w}(n-1) + \eta\,\varphi(\mathbf{u}(n))e(n) \tag{2.41}$$

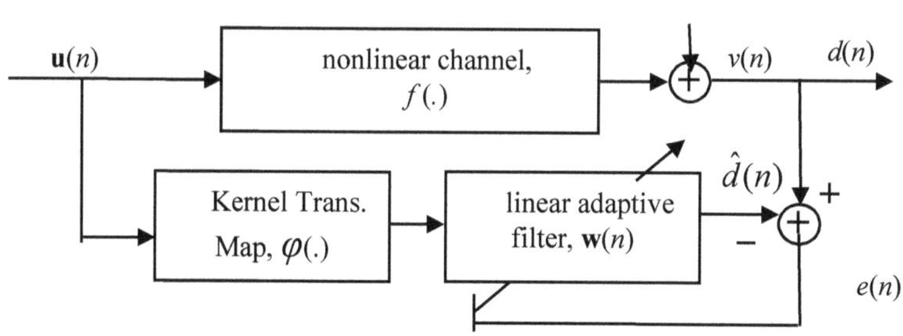


**Fig. 2.2** Kernel adaptive filter for system identification

The dimension size of $\varphi(\mathbf{u}(n))$ may be large, increasing computation. Therefore, an alternative method of performing the algorithm may be used. If we assume the initial weights are $\mathbf{w}(0) = 0$, then the weight update of (2.41) can be expressed as

$$\begin{aligned}
\mathbf{w}(n) &= \mathbf{w}(n-1) + \eta\,\varphi(\mathbf{u}(n))e(n) \\
&= \underbrace{[\mathbf{w}(n-2) + \eta\,\varphi(\mathbf{u}(n-1))e(n-1)]}_{\mathbf{w}(n-1)} + \eta\,\varphi(\mathbf{u}(n))e(n) \\
&= \mathbf{w}(n-2) + \eta\,[\varphi(\mathbf{u}(n-1))e(n-1) + \varphi(\mathbf{u}(n))e(n)] \\
&\ \ \vdots \\
&= \mathbf{w}(0) + \eta \sum_{i=1}^{n} e(i)\varphi(\mathbf{u}(i)) \\
&= \eta \sum_{i=1}^{n} e(i)\varphi(\mathbf{u}(i))
\end{aligned} \tag{2.42}$$

Note that the weight estimate $\mathbf{w}(n)$ is a linear combination of the previous and current transformed inputs, where the $e(n)$ prediction errors are the combining weights (after scaling with $\eta$).

Finally, we can use this representation of the weights to find the output of the system at iteration $n$, which is

$$\begin{aligned}
\mathbf{w}(n-1)^T\varphi(\mathbf{u}(n)) &= \left[\eta \sum_{i=1}^{n-1} e(i)\varphi(\mathbf{u}(i))\right]^T \varphi(\mathbf{u}(n)) \\
&= \eta \sum_{i=1}^{n-1} e(i)\varphi(\mathbf{u}(i))^T\varphi(\mathbf{u}(n)) \\
&= \eta \sum_{i=1}^{n-1} e(i)\kappa(\mathbf{u}(i), \mathbf{u}(n))
\end{aligned} \tag{2.43}$$

where $\kappa(\ .\ ,\ \ .\ )$ is used for the inner product.

Forming the algorithm in terms of inner products allows the use of kernel functions, and substantially reduces computation compared to directly mapping the input to the transform space using $\varphi(.)$.

A summary of the algorithm is shown in Fig. 2.3 (reproduced from [1]). Here, the notation $f_i$ represents the estimate of the nonlinear channel, and $i$ denotes iteration.

Note that KLMS requires $O(n)$ operations per iteration (where $n$ is the iteration), compared to $O(M)$ for LMS (where $M$ is the linear filter length). However, considering that the filter length may be infinite-dimensional for KLMS (for example, using the Gaussian kernel), the approach allows for practical implementation in this case.

Other considerations for KLMS include the selection of the parameters used. These parameters are kernel parameters $\sigma$, $q$ and step-size $\eta$. Also important is the

---

**Algorithm** The kernel least-mean-square algorithm

---

*Initialization*

choose step-size parameter $\eta$ and kernel $\kappa$

$\mathbf{a}_1(1) = \eta d(1), C(1) = \{\mathbf{u}(1)\}, f_1 = \mathbf{a}_1(1)\kappa(\mathbf{u}(1), \cdot)$

*Computation*

**while** $\{\mathbf{u}(i), d(i)\}$ available **do**

%compute the output

$$f_{i-1}(\mathbf{u}(i)) = \sum_{j=1}^{i-1} \mathbf{a}_j(i-1)\kappa(\mathbf{u}(i), \mathbf{u}(j))$$

%compute the error

$e(i) = d(i) - f_{i-1}(\mathbf{u}(i))$

%store the new center

$C(i) = \{C(i-1), \mathbf{u}(i)\}$

%compute and store the coefficient

$\mathbf{a}_i(i) = \eta e(i)$

**end while**

---

**Fig. 2.3** Summary of the KLMS algorithm

memory requirement of storing the previous $n$-1 inputs for evaluating the output. These areas are considered at various locations in this book (Chaps. 7–11).

## References


1. W. Liu, J. Principe, S. Haykin, *Kernel adaptive filtering : A comprehensive introduction* (Wiley Publishers, 2010)
2. S. Haykin, *Adaptive signal processing*, 4th ed. (Prentice-Hall, 2002)
3. P.S.R. Diniz, *Adaptive filtering: Algorithms and practical implementation*, 3rd ed. (Springer Publishers, 2008)
4. A.H. Sayed, *Fundamentals of adaptive filtering* (Wiley, New York, 2003)
5. S.L. Gay, J. Benesty, *Acoustic signal processing for telecommunication* (Kluwer, Boston, MA, 2000)
6. K. Ozeki, T. Umeda, "An adaptive filtering algorithm using an orthogonal projection to an affine subspace and its properties." Electron. Commun. Japan., **67**-A(5), 19–27 (1984)
7. S.G. Sankaran, A.A. (Louis) Beex, "Convergence behavior of affine projection algorithms." IEEE Trans. Signal Process., **48**, 1086–1096 (2000)
8. H.-C. Shin, A.H. Sayed, Mean-square performance of a family of affine projection algorithms. IEEE Trans. Signal Processing **52**, 90–102 (2004)

9. J.C. Principe, N.R. Euliano, W.C. Lefebvre, *Neural and adaptive systems: Fundamentals through simulations* (John Wiley & Sons, Inc., 2000)
10. S. Haykin, *Neural networks and learning machines*, 3rd ed. (Prentice-Hall, 2009)
11. J. Mathews, G. Sicuranza, *Polynomial signal processing* (Wiley Publishers, 2000)
12. I. Steinwart, D. Hush, C. Scovel, "An explicit description of the reproducing Kernel Hilbert Spaces of Gaussian RBF kernels". IEEE Trans. Inf. Theory, **52**(10), (2006)
13. J. Kivinen, A.J. Smola, R.C. Williamson, Online learning with kernels. IEEE Trans. Signal Process. **52**(8), 2165–2176 (2004)
14. N. Aronszajn, Theory of reproducing kernels. Trans. Amer. Math. Soc. **68**(3), 337–404 (1950)
15. S. Theodoridis, K. Slavakis, I. Yamada, Adaptive learning in a world of projections: A unifying framework for linear and nonlinear classification and regression tasks. IEEE Signal Process. Mag. **28**(1), 97–123 (2011)
16. C. Richard, J.C.M. Bermudez, P. Honeine, Online Prediction of time series data with Kernels. IEEE Trans. Signal Process. **57**(3), 1058–1066 (2009)
17. Y. Engel, S. Mannor, R. Meir, Kernel recursive least squares. IEEE Trans. Signal Process. **52**(8), 2275–2285 (2004)
18. W. Liu, P. Pokharel, J. Principe, "The Kernel least-mean-square algorithm". IEEE Trans. Signal Process., **56**(2), (2008)

# Chapter 3
# Complex Data for Adaptive Filters

This chapter describes how adaptive filters can be designed for use with complex data. The difficulty of determining the gradient of complex functions is considered, as well as Wirtinger's (or CR) Calculus approach for finding gradients of these functions. A description of the use of adaptive filtering with complex data is then provided, including the idea of widely linear estimation for complex data.

## 3.1 Complex Functions and Their Gradients

To understand the design of adaptive filters for complex-valued data, it is important to initially understand complex gradients. We consider this area in this section.

For the complex gradient of some function $f$, we can consider the gradient in terms of its components. The gradient operation may be considered a directional derivative for this case.

The gradient $\nabla f$ can be obtained from the derivative of $f$ in terms of each of its independent direction components. For the case of complex data, these are the real and imaginary directions. The "partial derivatives" are combined, forming a direction vector known as the gradient.

For example, we consider evaluating the complex gradient $\nabla_z f(z)$, or the gradient of function $f(z)$ with respect to complex scalar $z$. A simple example is

$$f(z) = az \tag{3.1}$$

where $a$ is a complex scalar value.

To illustrate the gradient, we write: $z = x + iy$, where $x$ is the real component, and $y$ is the imaginary component. Therefore,



T. K. Paul and T. Ogunfunmi, *Adaptive Filter Algorithms For Complex and Quaternion Data*, https://doi.org/10.1007/978-3-032-23798-9_3

$$\begin{aligned} f(x+iy) &= a \cdot (x+iy) \\ &= ax + iay \end{aligned} \tag{3.2}$$

where $a$ is a complex variable, and $x$ and $y$ are real-valued. We see here that the same complex scalar is applied to both the real and imaginary directions, specifically scalar $a$.

For the standard approach for finding the complex derivative (known as the *standard complex derivative*), finding the complex gradient of $f(.)$ is based on the idea that the derivative should be the same regardless of the direction in which the gradient is evaluated. In other words, the derivative should be the same whether travelling across the real axis or imaginary axis.

Considering the definition of the standard complex derivative [1]:

$$\lim_{\Delta z \to 0} \frac{f(z+\Delta z) - f(z)}{\Delta z} = \frac{\partial f(z)}{\partial z} \tag{3.3}$$

where $\Delta z = \Delta x + i\Delta y$, the following derivatives should be equal to each other:

$$\lim_{\Delta x \to 0} \frac{f(x+iy+\Delta x) - f(x+iy)}{\Delta x} = \frac{\partial f(x+iy)}{\partial x}$$

$$\lim_{\Delta y \to 0} \frac{f(x+iy+i\Delta y) - f(x+iy)}{i\Delta y} = \frac{1}{i} \cdot \frac{\partial f(x+iy)}{\partial y}$$

Therefore,

$$\frac{\partial f(x+iy)}{\partial x} = \frac{1}{i} \cdot \frac{\partial f(x+iy)}{\partial y} \tag{3.4}$$

For an arbitrary function, $f(x+iy) = u + iv$, we can say

$$\begin{aligned} &\frac{\partial [u+iv]}{\partial x} = \frac{1}{i} \cdot \frac{\partial [u+iv]}{\partial y} \\ &\frac{\partial u}{\partial x} + i\frac{\partial v}{\partial x} = \frac{\partial v}{\partial y} - i\frac{\partial u}{\partial y} \Rightarrow \boxed{\therefore \frac{\partial u}{\partial x} = \frac{\partial v}{\partial y} \, and \, \frac{\partial v}{\partial x} = \frac{-\partial u}{\partial y}} \end{aligned} \tag{3.5}$$

These are known as the Cauchy-Riemann equations.

For finding the gradient, these equations are required to be satisfied for the derivative to be considered to exist, based on the standard complex derivative [1]. It can also be shown that these conditions are sufficient to ensure the same derivative occurs in all directions toward $z$ [2].

Considering Eq. (3.2), based on (3.4), the partial derivatives are

$$\frac{\partial f(x+iy)}{\partial x} = a, \ \frac{1}{i} \cdot \frac{\partial f(x+iy)}{\partial y} = a \tag{3.6}$$

Here, Eq. (3.4) is satisfied, and the derivative is said to exist using the standard complex derivative. Therefore, the complex derivative result, across both the real and imaginary axes, is $a$.

However, not all complex functions have a derivative based on this approach. For example, consider again the complex gradient $\nabla_z f(z)$ (i.e. gradient of $f(z)$ w.r.t. $z$), although now for the function:

$$f(z) = bz^* \tag{3.7}$$

For this case

$$\begin{aligned} f(x+iy) &= b \cdot (x+iy)^* \\ &= bx - iby \end{aligned} \tag{3.8}$$

and, considering (3.4):

$$\frac{\partial f(x+iy)}{\partial x} = b, \frac{1}{i} \cdot \frac{\partial f(x+iy)}{\partial y} = -b \tag{3.9}$$

Equation (3.4) is not satisfied here since the derivative is different across the real and imaginary axes. The derivative is said not to exist in this case, based on the standard complex derivative.

The gradient examples described here (based on the standard complex gradient) indicate that the complex gradient may be seen not to strictly exist for some functions, for example, $f(z)$ in (3.7).

In order to evaluate gradients of arbitrary complex functions, an approach known as Wirtinger's Calculus (or $\mathbb{CR}$ Calculus) was developed. This calculus will be seen to be useful in evaluating cost function gradients for designing complex adaptive filters.

## 3.2 $\mathbb{CR}$ (or Wirtinger's) Calculus

Wirtinger's Calculus is an approach that allows the derivative of any complex function, $f(z)$, to be obtained, provided the following. Consider $f(z)$ expressed as

$$\begin{aligned} f(x+iy) &= f(x,y) \\ &= u(x,y) + iv(x,y) \end{aligned} \tag{3.10}$$

where $u(x, y)$ and $v(x, y)$ are the real, imaginary components of the function $f(x + iy)$ (both real-valued functions). We can find the derivative of $f(x + iy)$ using Wirtinger's Calculus if both $u(x, y)$ and $v(x, y)$ are real differentiable with respect to

$x$ and $y$. This is a less stringent requirement than the Cauchy-Riemann equations for the standard complex derivative.

For Wirtinger's Calculus, the following rules were derived for finding the derivative with respect to $z$ and $z^*$, called the *Wirtinger's derivative* and *conjugate Wirtinger's derivative* [3]:

$$\frac{\partial f(z)}{\partial z} = \frac{1}{2}\left(\frac{\partial f(x+iy)}{\partial x} - i\frac{\partial f(x+iy)}{\partial y}\right)$$

$$\frac{\partial f(z)}{\partial z^*} = \frac{1}{2}\left(\frac{\partial f(x+iy)}{\partial x} + i\frac{\partial f(x+iy)}{\partial y}\right) \tag{3.11}$$

These can be compared to the standard complex derivative results:

$$\frac{\partial f(z)}{\partial z} = \frac{\partial f(x+iy)}{\partial x}\left(= -i\frac{\partial f(x+iy)}{\partial y} \text{ for derivative to exist}\right)$$

$$\frac{\partial f(z)}{\partial z^*} = \frac{\partial f(x+iy)}{\partial x}\left(= i\frac{\partial f(x+iy)}{\partial y} \text{ for derivative to exist}\right) \tag{3.12}$$

When the standard complex derivative exists, Wirtinger's Calculus gives the same result. However, Wirtinger's Calculus can be applied to any complex function $f(z)$ provided its real and imaginary components are real differentiable as described previously.

For $f(z) = bz^*$, whose derivative with respect to $z$ is considered not to exist with the standard complex derivative (see (3.9)), the use of Wirtinger's Calculus yields.

$$\frac{\partial f(z)}{\partial z} = \frac{1}{2}(b - b) = 0 \tag{3.13}$$

Specifically, from Wirtinger's Calculus:

$$\frac{\partial z}{\partial z} = \frac{\partial z^*}{\partial z^*} = 1, \ \frac{\partial z}{\partial z^*} = \frac{\partial z^*}{\partial z} = 0 \tag{3.14}$$

Finally, it can be shown that any complex function $f(z)$ whose real, imaginary components are real differentiable (with respect to $x$, $y$) can be written in the form:

$$f(z) = f\left(z, z^*\right) \tag{3.15}$$

where in this form $f(z)$ is referred to as *analytic in* $z$ when $z^*$ is fixed, and *analytic in* $z^*$ when $z$ is fixed [1]. A function being analytic on some domain (in this case, $z$) refers to it being expressed as a Taylor series on that domain.

Any function that is *analytic in* $z$ is complex differentiable w.r.t. $z$ (based on the standard complex derivative), while any function that is *analytic in* $z^*$ is complex differentiable w.r.t. $z^*$. Wirtinger's Calculus allows functions which are *nonanalytic in* $z$ or *nonanalytic in* $z^*$ also to be differentiated by considering $z^*$ fixed when differentiating w.r.t. $z$, and vice-versa.

Therefore, Wirtinger's Calculus is an efficient approach to differentiate any complex function $f(z)$ whose real, imaginary components are real differentiable. When $f(z)$ is analytic in the variable of differentiation, Wirtinger's Calculus result coincides with the standard complex derivative. Wirtinger's Calculus can also be seen to be less effort than performing real differentiation with respect to the real and imaginary components of $f(z)$ (where $z = x + iy$), and recombining the results to find a result for $d[f(z)]/dz$ in terms of $z$ instead of $x$ and $y$.

## 3.3 Complex Adaptive Filtering

The use of Wirtinger's Calculus can be seen to enable the design of complex adaptive filters. The cost functions for complex adaptive filters are typically real-valued functions. For example, the cost function for the complex LMS is $J(n) = |e(n)|^2$. The use of Wirtinger's Calculus is important because the standard complex gradient does not exist for these functions. Here, we illustrate the design of complex adaptive filters by describing the derivation of the complex LMS algorithm.

### 3.3.1 *The Complex LMS Algorithm*

The complex version of the LMS algorithm is essentially similar to the real-valued version, aside from the fact that the system inputs, filter weights are complex. The goal of the algorithm is finding the complex filter $\mathbf{w}(n)$ to minimize the squared error magnitude $|e(n)|^2$, where $e(n) = d(n) - \hat{d}(n)$, with $d(n)$ being desired response, and $\hat{d}(n) = \mathbf{w}^H(n)\mathbf{u}(n)$ being the estimator output, based on input vector $\mathbf{u}(n)$. All signals used in the system (i.e. $\mathbf{w}(n)$, $\mathbf{u}(n)$, $d(n)$, $\hat{d}(n)$, $e(n)$) are complex-valued.

Based on gradient descent, the complex LMS update, similar to LMS, is

$$\mathbf{w}(n) = \mathbf{w}(n-1) - \mu \nabla_{\mathbf{w}*} J(n) \tag{3.16}$$

where $\nabla_{w*}$ denotes gradient with respect to $\mathbf{w}^*(n)$, $\mu$ is the step-size parameter, and complex function $J(n) = |e(n)|^2$ is the cost function.

Using Wirtinger's Calculus, the gradient can be seen to be

$$\begin{aligned}\nabla_{\mathbf{w}*} J(n) &= \nabla_{\mathbf{w}*}\{|e(n)|^2\} \\ &= \nabla_{\mathbf{w}*}\left\{e(n)e^*(n)\right\}\end{aligned}$$

$$= \nabla_{\mathbf{w}*}\big\{\big(d(n) - \mathbf{w}^H(n)\mathbf{u}(n)\big)\big(d^*(n) - \mathbf{w}^T(n)\mathbf{u}^*(n)\big)\big\}$$
$$= -\mathbf{u}(n)\big(d^*(n) - \mathbf{w}^T(n)\mathbf{u}^*(n)\big)$$
$$= -\mathbf{u}(n)e^*(n) \tag{3.17}$$

Note that $\nabla_{w*}J(n)$ can be shown not to exist using the standard complex gradient here.

Therefore, the weight update for the complex LMS algorithm is

$$\mathbf{w}(n) = \mathbf{w}(n-1) + \mu\mathbf{u}(n)e^*(n) \tag{3.18}$$

and the complex LMS algorithm (for every iteration $n$) is summarized as

$$\hat{d}(n) = \mathbf{w}^H(n-1)\mathbf{u}(n) \quad \textbf{(filter output)}$$
$$\mathbf{w}(n) = \mathbf{w}(n-1) - \mu\mathbf{u}(n)e^*(n) \quad \textbf{(filter update)}$$
$$e(n) = d(n) - \hat{d}(n) \quad \textbf{(filter error)}$$

The algorithm can be seen to be essentially the same as the standard LMS (aside from conjugation terms), and is equivalent to LMS when real-valued data is used.

However, the estimation form $\hat{d}(n) = \mathbf{w}^H(n)\mathbf{u}(n)$ has been observed not to provide complete statistical information for regression with complex data. An alternative estimation form is described in the next section, which improves estimation performance with complex data.

## 3.4 Linear Versus Widely Linear Estimation

Widely linear (WL) estimation, and its benefits for adaptive filtering, are considered in references [4–8]. Here we describe the approach.

WL estimation allows for improved mean-square estimation (MS) performance when using complex data. For mean-square estimation, the problem can be described as estimating a scalar random variable $y$, in terms of an observed random vector, $\mathbf{x}$. The estimate $\hat{y}$, minimizing mean-square error (MSE) is the regression based on conditional expectation, $E[y|\mathbf{x}]$.

When $\mathbf{x}$ and $y$ are jointly Gaussian and real-valued, the regression is linear in $\mathbf{x}$. However, with complex-valued data, the regression can be seen to be linear in $\mathbf{x}$ as well as $\mathbf{x}^*$, and is referred to as widely linear (WL) [5].

For linear MSE (LMSE), the estimate has the form:

$$y = \mathbf{h}^H\mathbf{x} \tag{3.19}$$

where $\mathbf{h}$ is the optimum linear weight vector for estimation, and $^H$ denotes the Hermitian transpose.

For the widely linear MSE (WLMSE):

$$y' = \mathbf{h}^H \mathbf{x} + \mathbf{g}^H \mathbf{x}^* \tag{3.20}$$

Equation (3.20) is the general form for the regression with complex data. From (3.20), we see estimate $y'$ is not a linear function of $\mathbf{x}$. However, the order $k$ moments for $y'$ are completely defined from order $k$ moments of $\mathbf{x}$ and $\mathbf{x}^*$, forming a type of linearity. For this reason, (3.20) is called a wide-sense linear filter [5].

This approach can be seen to utilize the full second-order statistics of complex input $\mathbf{x}$, which can be described from two covariance matrices. The covariance matrix:

$$\boldsymbol{C} = E\left[\boldsymbol{x}\boldsymbol{x}^H\right] \tag{3.21}$$

and the pseudo-covariance matrix (also referred to as the complementary covariance matrix) [6]:

$$\boldsymbol{P} = E\left[\boldsymbol{x}\boldsymbol{x}^T\right]. \tag{3.22}$$

These statistical quantities can be seen to vary based on input circularity, where circularity is a signal characteristic affecting performance. In general, the circularity of a complex random vector, $\mathbf{x}$, is dependent on whether its pdf is rotation invariant. If $\mathbf{x}$ and $\mathbf{x}e^{j\theta}$ have the same pdf for all $\theta$, the pdf of $\mathbf{x}$ is considered rotation invariant. In this case, the vector $\mathbf{x}$ is referred to as circular in the strict sense [6], or simply circular.

Alternatively, if complex vector $\mathbf{x}$ is zero-mean and $E[\mathbf{x}\mathbf{x}^T] = \mathbf{0}$, then $\mathbf{x}$ is said to be second-order circular (or proper). Otherwise, $\mathbf{x}$ is considered noncircular when $E[\mathbf{x}\mathbf{x}^T] \neq \mathbf{0}$. The circularity of the complex input for MS estimation is known to affect the performance for LMSE [4–9].

The WLMSE approach improves MSE performance compared to LMSE based on the use of complete second-order input statistics. The approach is considered in Chap. 8 for improving the performance of kernel methods for complex data.

## References

1. K. Kreutz-Delgado, "The complex gradient operator and the CR-calculus." http://citeseerx.ist.psu.edu/viewdoc/download?doi=10.1.1.86.6515&rep=rep1&type=pdf
2. S. Haykin, *Adaptive signal processing*, 4th ed., published by Prentice-Hall, (2002)
3. P. Bouboulis, S. Theodoridis, "Extension of Wirtinger's calculus to reproducing Kernel Hilbert spaces and the complex Kernel LMS. IEEE Trans. Signal Process., **59**(3), (2011)
4. D.P. Mandic, S.L. Goh, *Complex valued nonlinear adaptive filters: Noncircularity* (Wiley, Widely Linear and Neural Models, 2009)
5. B. Picinbono, Bernard, Chevalier, Pascal, "Widely linear estimation with complex data. IEEE Trans. Signal Process., **43**(8), 2030–2033 (1995)
6. T. Adali, H. Li, R. Aloyious, On properties of the widely linear MSE filter and its LMS implementation. in *Proc Conf Inf Sci Syst (CISS), Baltimore, MD* (2009)

7. T. Adali, S. Haykin, *Complex-valued adaptive signal processing*, ser. Adaptive Signal Processing: Next Generation Solutions (Wiley, Hoboken, NJ, 2010)
8. Y. Xia, C.C. Took, D.P. Mandic, An augmented affine projection algorithm for the filtering of noncircular complex signals. Signal Process. **90**(6), 1788–1799 (2010)
9. B. Picinbono, On circularity. IEEE Trans. Signal Process. **42**(12), 3473–3482 (1994)

# Chapter 4
# The Complex Kernel Least Mean Square Algorithm

This chapter describes the Complex Kernel LMS (CKLMS) method by Bouboulis and Theodoridis [1, 2]. The description here begins with providing definitions of reproducing kernels for complex data. A modified form of Wirtinger's Calculus suitable for finding gradients of complex functions on Hilbert spaces was developed in [1], which is also summarized here. These areas are then used to derive the algorithm. The theory presented describes requirements for designing kernel adaptive filters with complex data, and forms the foundation for much of the work in later chapters.

## 4.1 Theory of Reproducing Kernels for Complex Data

Background theory on kernel functions for complex data and their associated Reproducing Kernel Hilbert Spaces (RKHS) is summarized in this section. Additional details on this area are provided with references [1, 2].

The kernel function and RKHS are associated with a general field $F$ , which may be a real field (denoted $R$ ) or complex (denoted $C$ ). Given a function $\kappa : X \times X \rightarrow F$ with possible inputs $x_1, \ldots, x_N \in X$ , we may define an $N$ x $N$ matrix **K** with elements $K_{i,j} = \kappa\left(x_i, x_j\right)$ , for $i, j = 1, \ldots, N$, called the Gram (or Kernel) matrix of $\kappa$ with respect to $x_1, \ldots, x_N$.

A Hermitian matrix **K** can be referred to as positive definite if

$$c^H Kc = \sum_{i=1,j=1}^{N,N} c_i^* c_j K_{i,j} \geq 0 \tag{4.1}$$

for all $c_i \in F$ , $i = 1, \ldots, N$, where $^*$ denotes conjugation. A function $\kappa$ satisfying (4.1) is known as a positive definite kernel, or simply as a kernel. Although actually



T. K. Paul and T. Ogunfunmi, *Adaptive Filter Algorithms For Complex and Quaternion Data*, https://doi.org/10.1007/978-3-032-23798-9_4

positive semi-definite, the notation "positive definite kernel" is used since the former term is cumbersome, and the distinction is unnecessary for the kernel definition [1].

It has been shown that for every positive definite kernel $\kappa$, there exists a single class $H$ of functions (i.e. functions $f$ defined to act on set $X$), where $H$ is a Hilbert space with a unique inner product, and where $\kappa$ can reproduce the entire space of functions, $H$ [1]. A Hilbert space $H$ may be described as a space where an inner product $< \cdot, \cdot >_H$ can be defined with a corresponding norm $\|\cdot\|_H$, and where $H$ is considered complete with respect to the norm. Therefore, $\kappa$ may also be referred to as a reproducing kernel, and $H$ its associated Reproducing Kernel Hilbert Space (RKHS). For the complex case, $H$ is a class of complex-valued functions, referred to as a complex RKHS.

An RKHS and its associated kernel have the properties:

(1) For every $x \in X$, $\kappa(\cdot, x)$ belongs to $H$.
(2) $\kappa$ has the so-called "reproducing property", i.e.

$$f(x) = < f, \kappa(\cdot, x) >_H, \quad \textit{for all } f \in H$$

and, in particular $\kappa(x, y) = < \kappa(\cdot, y), \kappa(\cdot, x) >_H$.

Additionally, we can define: $\Phi : X \to H : \Phi(x) = \kappa(\cdot, x)$, referred to as the feature map of $H$. Also, for complex Hilbert spaces (i.e. where $F = C$), the inner product is sesquilinear and Hermitian, i.e.

$$\begin{aligned} < af + bg, h >_H &= a < f, h >_H + b < g, h >_H \\ < f, ag + bh >_H &= a^* < f, g >_H + b^* < f, h >_H \\ < f, g >_H^* &= < g, f >_H \end{aligned} \tag{4.2}$$

for all $f, g, h \in H$ and $a, b \in C$.

As an alternative to complex RKHS, which requires the use of less established complex kernels, reference [1] introduces the notion of "complexified" RKHS spaces. The idea is to form an RKHS created from real RKHS spaces, where the new space possesses a suitable complex structure for defining a complex inner product and reproducing kernel. The derivation of the complexified RKHS is described here.

Define $X \subseteq R^v$, $X^2 = X \times X \subseteq R^v$ and $X = \{x + iy; x, y \in X\} \subseteq C^v$ possessing a complex product structure. Now let $H$ be a real RKHS with real kernel $\kappa : X^2 \times X^2$ and inner product $< \cdot, \cdot >_H$. Then each $f \in H$ can be considered a function defined on either. $X^2$ or $X$, i.e. $f(\mathbf{z}) = f(\mathbf{x} + i\mathbf{y}) = f(\mathbf{x}, \mathbf{y})$.

Now define $H^2 = H \times H$. It can be verified $H^2$ is a Hilbert space with inner product:

$$< f, g >_{H^2} = < f_1, g_1 >_H + < f_2, g_2 >_H \tag{4.3}$$

where $f = (f_1, f_2)$, $g = (g_1, g_2)$.

In order to provide $H^2$ a complex structure, a space is defined: $\mathbb{H} = \{f = f_1 + i * f_2;\ f_1, f_2 \in H\}$ in [1], along with inner product:

$$<f, g>_{\mathbb{H}} = <f_1, g_1>_H + <f_2, g_2>_H + i <f_2, g_1>_H - i <f_1, g_2>_H \quad (4.4)$$

where $f = f_1 + if_2$ , $g = g_1 + ig_2$ . Therefore $f,\ g : X \subseteq C^v \to C$ . The space $\mathbb{H}$ can be verified to be a complex RKHS with associated kernel $\kappa$ from before [1]. The associated kernel is real here. $\mathbb{H}$ is referred to as the complexification of $H$.

To provide a mapping for complex input data to the complexified RKHS $\mathbb{H}$, the following rule is used:

$$\Phi(z) = \Phi(x + iy) = \Phi(x, y) = \Phi(x, y) + i \cdot \Phi(x, y) \quad (4.5)$$

where $\Phi$ is the mapping for complex input $z = x + iy$, and $\Phi$ is the feature map of the real kernel $\kappa$, i.e. $\Phi(x, y) = \kappa(\cdot, (x, y))$. Note $\Phi$ does not represent the feature map associated with the complex RKHS for $\mathbb{H}$ . A complex inner product for $\mathbb{H}$ can then be expressed as

$$\begin{aligned} < \Phi(z), \Phi(z') >_{\mathbb{H}} &= 2 < \Phi(x, y), \Phi(x', y') >_H \\ &= 2\kappa\left((x, y), (x', y')\right) \end{aligned} \quad (4.6)$$

The obtained result, referred to in [1] as the complexified real kernel trick, shows that any algorithm formed from complex inner products (i.e. $\mathbf{w}^H\mathbf{z}$, where $\mathbf{z} = x + iy$, $\mathbf{w} = u + iv$) can alternatively be formed, based on the approach here, by replacing each inner product with a positive definite real kernel, whose arguments are the components of inputs $\mathbf{z}$ and $\mathbf{w}$ (i.e. $\kappa((x, y), (u, v))$). This approach provides an alternative to the use of complex kernels, which are currently less established in machine learning.

The complex and "complexified" kernel functions and associated RKHS definitions formed the basis for new algorithms derived in later chapters.

## 4.2 Modified Wirtinger's Calculus for Hilbert Spaces

Wirtinger's Calculus, described in Chap. 3, shows an approach for finding the gradient of functions on complex data. However, for adaptive algorithms using the complex kernels presented, we need to find the gradient of functions defined on complex Hilbert spaces (specifically, defined on complex RKHS).

A modified form of Wirtinger's Calculus (WC) generalized for Hilbert spaces was obtained in reference [1], and is summarized here. The calculus should apply to general Hilbert spaces, and is obtained using a Fréchet derivative.

### 4.2.1 Fréchet Derivative

The Fréchet derivative generalizes the notion of differentiability to Banach or Hilbert spaces, and is described here. Consider a Hilbert space $H$ over a general field $F$ (which may be real $R$ or complex $C$ ). An operator $T : H \rightarrow F^{\nu}$ may be referred to as Fréchet differentiable at $f_0$ if a linear continuous operator $W = (W_1, W_2, \ldots, W_{\nu}) : H \rightarrow F^{\nu}$ exists such that

$$\lim_{\|h\|_H \to 0} = \frac{\|T(f_0 + h) - T(f_0) - W(h)\|_{F^{\nu}}}{\|h\|_H} = 0 \tag{4.7}$$

where $\|.\|_H = \sqrt{\langle ., . \rangle_H}$ is the induced norm, $\langle ., . \rangle_H$ the dot product for the Hilbert space. $\boldsymbol{W}$ is then called the Fréchet derivative, and can be denoted $d\boldsymbol{T}(f_0)$: $H \rightarrow F^{\nu}$.

From [1], some specific cases of Fréchet derivative (4.7) are as follows. When $\nu = 1$, $\boldsymbol{W}$ may be replaced with an inner product, i.e. $W(h) = < h, w >_H$ where $w^*$ is called the gradient of $T$ w.r.t. $\boldsymbol{f}$ at $f_0$, and is denoted $w^* = \nabla T(f_0)$.

When $\nu > 1$, we have vector $\boldsymbol{T} = (T_1, T_2, \ldots, T_{\nu})$, and the gradient can be seen to be

$$\partial T(f_0)(h) = \begin{pmatrix} \langle h, \nabla T_1(f_0)^* \rangle_H \\ \vdots \\ \langle h, \nabla T_{\nu}(f_0)^* \rangle_H \end{pmatrix} \tag{4.8}$$

Finally, for representing partial derivatives, we can define operator $T : H^{\mu} \rightarrow F$ on Hilbert space $H^{\mu}$, with inner product.

$$\langle f, g \rangle_{H^{\mu}} = \sum_{\iota=1}^{\mu} \langle f_{\iota}, g_{\iota} \rangle_H \tag{4.9}$$

where $\boldsymbol{f} = (f_1, f_2, \ldots, f_{\mu})$, $\boldsymbol{g} = (g_1, g_2, \ldots, g_{\mu})$ . $T(\boldsymbol{f})$ is considered Fréchet differentiable at $\boldsymbol{f}_0$ w.r.t. $f_{\iota}$ iff there exists $w \in H$ so that

$$\lim_{\|h\|_H \to 0} = \frac{T(f_0 + [h]_{\iota}) - T(f_0) - \langle [h]_{\iota}, w \rangle_H}{\|h\|_H} = 0 \tag{4.10}$$

where $[h]_{\iota} = (0, 0, \ldots, 0, h, 0, \ldots, 0)$ is the element of $H^{\mu}$ with zero entries everywhere except at place $\iota$, and $w^*$ is the gradient of $T$ at $\boldsymbol{f}_0$, denoted $w^* = \nabla_{\iota} T(f_0)$. The Fréchet partial derivative at $\boldsymbol{f}_0$ w.r.t. $f_{\iota}$ can be written

$$\frac{\partial T}{\partial f_{\iota}}(f_0)(h) = \langle [h]_{\iota}, w \rangle_H \tag{4.11}$$

### 4.2.2 Derivation of Modified Calculus

We now describe the Fréchet-based differentiation rules from [1] applicable to complex functions defined on $\mathbb{H}$, the Hilbert space for complex data described in Sect. 4.1. These rules resemble Wirtinger's Calculus described in Chap. 3, and are obtained based on the isomorphism between $\mathbb{H}$ and the real space $H^2$ described earlier. The modified WC shown here was used for performing the gradient operation for deriving kernel algorithms for complex-valued data in Chap. 7, and also for finding a similar calculus for quaternions in Chap. 10.

Consider function $\boldsymbol{T}$ defined on an open subset $A$ of $H$ , i.e. $T : A \subseteq H \to C$ , where (based on the isomorphism between $\mathbb{H}$ , $H^2$).

$$T(f) = T\left(u_f + iv_f\right) = T_r\left(u_f, v_f\right) + iT_i\left(u_f, v_f\right) \tag{4.12}$$

with $u_f$, $v_f \in \mathbb{H}$ and $T_r$, $T_i$ real-valued functions defined on $H^2$ .

Since for the output, $C$ and $R^2$ are also isomorphic, the following notations may be seen equivalent:

$$T(f) = T_r\left(u_f, v_f\right) + iT_i\left(u_f, v_f\right)$$

$$T(f) = \left[T_r\left(u_f, v_f\right), T_i\left(u_f, v_f\right)\right] \tag{4.13}$$

The derivative of $\boldsymbol{T}$ is considered Fréchet complex differentiable at $c \in \mathbb{H}$ if there exists $w \in \mathbb{H}$ such that

$$\lim_{\|h\|_{\mathbb{H}} \to 0} = \frac{T(c+h) - T(c) - \langle h, w \rangle_{\mathbb{H}}}{\|h\|_{\mathbb{H}}} = 0 \tag{4.14}$$

where $\boldsymbol{w}^*$ is referred to as the Fréchet complex gradient of $\boldsymbol{T}$ at $\boldsymbol{c}$, and is denoted $w^* \equiv \nabla T(c)$ . The complex derivative is denoted $dT(c)(h) = \langle h, w \rangle_{\mathbb{H}}$ . The inner product $\langle h, w \rangle_{\mathbb{H}}$ in (4.14) places a great amount of structure on $\boldsymbol{T}$, and is seen to give rise to the strong properties of the complex derivative.

Additionally, it can be seen if $dT(c)$ exists, then $d^nT(c)$ also exists for all integer $n$. If $\boldsymbol{T}$ is differentiable at any point $\boldsymbol{c} \in A$, then $\boldsymbol{T}$ may be called Fréchet holomorphic in $A$, or Fréchet complex analytic in $A$, since it has a Taylor series expansion [1].

$$T(c+h) = \sum_{n=0}^{\infty} \frac{1}{n!} d^nT(c)(h, h, ..., h) \tag{4.15}$$

where $d^nT(c)(h, h, ..., h)$ denotes that multilinear map $d^nT(c)$ is applied to ($\boldsymbol{h}$, $\boldsymbol{h}$, ..., $\boldsymbol{h}$) (can also be expressed $d^nT(c)h^n$ ).

Some additional definitions on differentiability are:

- ***T*** is considered Fréchet differentiable in the complex sense if the complex derivative exists.
- ***T*** is considered Fréchet differentiable in the real sense if the real derivative exists when considered as the vector operator $T$: $\mathrm{H}^2 \rightarrow \mathrm{H}^2$
- ***T*** is considered Fréchet complex analytic at ***c*** if Fréchet holomorphic in the neighborhood around ***c***.
- ***T*** is considered Fréchet real when $T_r$, $T_i$ all have a Taylor series expansion in the real sense.

If $T_r$, $T_i$, are Fréchet differentiable in the real sense (regardless of analyticity of ***T***), differentiation rules were obtained in [1], similar to the WC of Chap. 3, applicable for complex functions defined on $\mathbb{H}$ .

Considering the first-order Taylor series expansions of $T_r$, $T_{i.}$

$$T_r(c+h) = T_r(c) + \langle h_1, \nabla_u T_r(c)\rangle_H + \langle h_2, \nabla_v T_r(c)\rangle_H + o(\|h\|_{H^2})$$

$$T_i(c+h) = T_i(c) + \langle h_1, \nabla_u T_i(c)\rangle_H + \langle h_2, \nabla_v T_i(c)\rangle_H + o(\|h\|_{H^2})$$

Multiplying the second equation by $i$, and adding to the first equation

$$\begin{aligned} T(c+h) = T(c) &+ \langle h_1, \nabla_u T_r(c) - i\nabla_u T_i(c)\rangle_{\mathbb{H}} \\ &+ \langle h_2, \nabla_v T_r(c) - i\nabla_v T_i(c)\rangle_{\mathbb{H}} + o(\|h\|_{\mathbb{H}}) \end{aligned}$$

If we define:

$$\nabla_u T(c) = \nabla_u T_r(c) + i\nabla_u T_i(c)\nabla_v T(c) = \nabla_v T_r(c) + i\nabla_v T_i(c)$$

Simplifying based on these results

$$T(c+h) = T(c) + \left\langle h_1, (\nabla_u T(c))^*\right\rangle_{\mathbb{H}} + \left\langle h_2, (\nabla_v T(c))^*\right\rangle_{\mathbb{H}} + o(\|h\|_{\mathbb{H}})$$

Using the following complex-valued properties $h_1 = \frac{1}{2}[h+h^*]$, $h_2 = \frac{-i}{2}[h-h^*]$, we can write

$$\begin{aligned} T(c+h) &= T(c) + \left\langle \frac{1}{2}[h+h^*], (\nabla_u T(c))^*\right\rangle_{\mathbb{H}} + \left\langle \frac{-i}{2}[h-h^*], (\nabla_v T(c))^*\right\rangle_{\mathbb{H}} + o(\|h\|_{\mathbb{H}}) \\ &= T(c) + \frac{1}{2}\langle h, (\nabla_u T(c) - i\nabla_v T(c))^*\rangle_{\mathbb{H}} + \frac{1}{2}\langle h^*, (\nabla_u T(c) + i\nabla_v T(c))^*\rangle_{\mathbb{H}} + o(\|h\|_{\mathbb{H}}) \end{aligned} \tag{4.16}$$

From this equation, we can see the Taylor series expansion expressed in terms of ***h*** and $\boldsymbol{h}^*$. The same process can be extended for higher order Taylor series [1].

Based on (4.16), the following definitions were obtained. The Fréchet Wirtinger's gradient (or W- gradient) is

$$\begin{aligned}\nabla_f T(c) &= \frac{1}{2}(\nabla_u T(c) - i\nabla_v T(c)) \\ &= \frac{1}{2}(\nabla_u T_r(c) + \nabla_v T_i(c)) + \frac{i}{2}(\nabla_u T_i(c) - \nabla_v T_r(c))\end{aligned} \tag{4.17}$$

and the Fréchet Wirtinger's derivative (or W-derivative) is $(\partial T/\partial f)(c) : \mathbb{H} \to \mathbb{C}$, where $(\partial T/\partial f)(c)(h) =< h, \nabla_f T(c)^* >_{\mathbb{H}}$. Additionally, the Fréchet conjugate Wirtinger's gradient (or CW- gradient) is:

$$\begin{aligned}\nabla_{f^*} T(c) &= \frac{1}{2}(\nabla_u T(c) + i\nabla_v T(c)) \\ &= \frac{1}{2}(\nabla_u T_r(c) - \nabla_v T_i(c)) + \frac{i}{2}(\nabla_u T_i(c) + \nabla_v T_r(c))\end{aligned} \tag{4.18}$$

and the Fréchet conjugate Wirtinger's derivative (or CW-derivative) is $(\partial T/\partial f^*)(c) : \mathbb{H} \to \mathbb{C}$, where $(\partial T/\partial f^*)(c)(h) =< h, \nabla_{f^*} T(c)^* >_{\mathbb{H}}$.

Based on these Fréchet-based gradients, some differentiation properties for functions defined on complex Hilbert spaces were shown to be:

1. If $\boldsymbol{T}(\boldsymbol{f})$ is $\boldsymbol{f}$-holomorphic at $\boldsymbol{c}$ (i.e. it has a Taylor series expansion with respect to $\boldsymbol{f}$ at $\boldsymbol{c}$), then its Fréchet W-derivative at $\boldsymbol{c}$ reverts to the standard Fréchet complex derivative, and its Fréchet CW-derivative is zero, i.e. $\nabla_{f^*} T(c) = 0$.
2. If $\boldsymbol{T}(\boldsymbol{f})$ is $\boldsymbol{f}^*$-holomorphic at $\boldsymbol{c}$ (i.e. it has a Taylor series expansion with respect to $\boldsymbol{f}^*$ at $\boldsymbol{c}$), then $\nabla_f T(c) = 0$.
3. If $T(f) =< h, f >_{\mathbb{H}}$, then $\nabla_{f^*} T(c) = h$ for every $\boldsymbol{c}$.
4. If $T(f) =< h, f^* >_{\mathbb{H}}$, then $\nabla_{f^*} T(c) = 0$ for every $\boldsymbol{c}$.
5. Product Rule: When $T, S : \mathbb{H} \to \mathbb{C}$ are Fréchet differentiable in the real sense at $c \in \mathbb{H}$, then:

$$\nabla_{f^*}(T \cdot S)(c) = \left[\nabla_{f^*} T(c)\right] S(c) + T(c)\left[\nabla_{f^*} S(c)\right]$$

The Fréchet-based definitions and properties described may be used for performing gradient operations for deriving kernel-based algorithms for complex data (see Chap. 7).

## 4.3 Derivation of Algorithm

This section summarizes the derivation of the Complex Kernel LMS (CKLMS) algorithm from [1]. The goal for the algorithm is to form filter $w$ to estimate the desired response $d(n)$, i.e. $\hat{d}(n) =< \Phi(z(n)), w >_S$, where $\mathbf{z}(n)$ is the complex input. $\Phi(z(n))$ facilitates the use of linear methods in the RKHS space for learning a nonlinear channel function, $f(z(n))$. The notation $\Phi(.)$ and inner product $< . >_S$ may refer to either a complex or complexified RKHS.

The goal of the complex KLMS algorithm is to minimize the average squared error, which can be written as

$$\begin{aligned} L(w) &= |e(n)|^2 = |d(n)- < \Phi(z(n)), w >_S|^2 \\ &= (d(n)- < \Phi(z(n)), w >_S)(d(n)- < \Phi(z(n)), w >_S)^* \\ &= (d(n)- < w^*, \Phi\big(z(n))^* >_S\big)(d^*(n)- < w, \Phi(z(n)) >_S) \end{aligned} \tag{4.19}$$

For deriving the algorithm, instantaneous error $L(w)$ is used instead of $E[L(w)]$, as described in Chap. 2. Using the modified Wirtinger's Calculus for finding the gradient of cost function $L(w)$ yields.

$$\nabla_{w*}L(w) = -e(n)^* \cdot \Phi(z(n)) \tag{4.20}$$

The weight update becomes

$$w(n) = w(n-1) - \mu\nabla_{w*}L(n)$$

$$w(n) = w(n-1) + \mu e(n)^* \cdot \Phi(z(n)) \tag{4.21}$$

For expressing the algorithm in terms of inner products, we follow the approach of Chap. 3 for KLMS. Assuming $w(0) = 0$, the weight equation becomes

$$\begin{aligned} w(1) &= \mu e(1)^* \cdot \Phi(z(1)) \\ w(2) &= \mu e(1)^* \cdot \Phi(z(1)) + \mu e(2)^* \cdot \Phi(z(2)) \\ &\vdots \\ w(n) &= \mu \sum_{k=1}^{n} e(k)^* \cdot \Phi(z(k)) \end{aligned} \tag{4.22}$$

The filter output can then be expressed as

$$\begin{aligned} \hat{d}(n) &=< \Phi(z(n)), w(n-1) >_S \\ &= \mu \sum_{k=1}^{n-1} e(k) < \Phi(z(n)), \Phi(z(k)) >_S \end{aligned} \tag{4.23}$$

where $e(n) = d(n) - \hat{d}(n)$ is the estimation error.

The use of the normalized algorithm form was also described in [1]. The weight update in this case is

$$w(n) = w(n) + \frac{\mu}{< \Phi(z(n)), \Phi(z(n)) >_S} e(n)^* \cdot \Phi(z(n)) \tag{4.24}$$

and the estimates (with the same process used for obtaining (4.23)) are

$$\begin{aligned}\hat{d}(n) &=< \Phi(z(n)), w(n-1) >_S \\ &= \mu \sum_{k=1}^{n-1} e(k) \frac{< \Phi(z(n)), \Phi(z(k)) >_S}{< \Phi(z(k)), \Phi(z(k)) >_S}\end{aligned} \tag{4.25}$$

Evaluating the inner product terms in (4.25) may be performed with either the complex or "complexified" kernel form described in Sect. 4.1. The normalized algorithm based on a "complexified" kernel is referred to as NCKLMS-1, while the other using pure complex kernels, is referred to as NCKLMS-2. Results for both kernel forms are shown in the following section.

## 4.4 Simulations

The simulations used to evaluate the performance of the NCKLMS-1, NCKLMS-2 algorithms are described here. The tests performed were for nonlinear channel equalization. The following two nonlinear channel types were evaluated: a "soft nonlinear channel" and a "strong nonlinear channel". Both consist of a linear portion followed by a memoryless nonlinearity.

The linear portion for both channels was

$$t(n) = (-0.9 + 0.8i) \cdot s(n) + (0.6 - 0.7i) \cdot s(n-1)$$

The memoryless nonlinearity for the soft nonlinear channel was

$$q(n) = t(n) + (0.1 + 0.15i) \cdot t^2(n) + (0.06 + 0.05i) \cdot t^3(n)$$

while for the strong nonlinear channel:

$$q(n) = t(n) + (0.2 + 0.25i) \cdot t^2(n) + (0.12 + 0.09i) \cdot t^3(n)$$

The model used is considered standard for nonlinear channel modeling tasks [1, 3, 4]. The signal at the receiver is corrupted by white Gaussian noise with a noise level of 16dB. The input signal generated used the form:

$$s(n) = 0.7\left(\sqrt{1-p^2}X(n) + i \cdot \rho Y(n)\right)$$

where $X(n)$, $Y(n)$ are Gaussian random variables. The value of $\rho$ used was either $\sqrt{2}/2$ for generating highly circular input, or 0.1 (where values close to either 0 or 1 generate highly non-circular input).

The performance of the NCKLMS-1 and NCKLMS-2 algorithms for performing nonlinear channel equalization is shown in Figs. 4.1 and 4.2.

The tests shown in this section are.

Fig. 4.1: NCKLMS-1, NCKLMS-2 for a soft nonlinear channel.

Fig. 4.2: NCKLMS-1, NCKLMS-2 for a strong nonlinear channel.

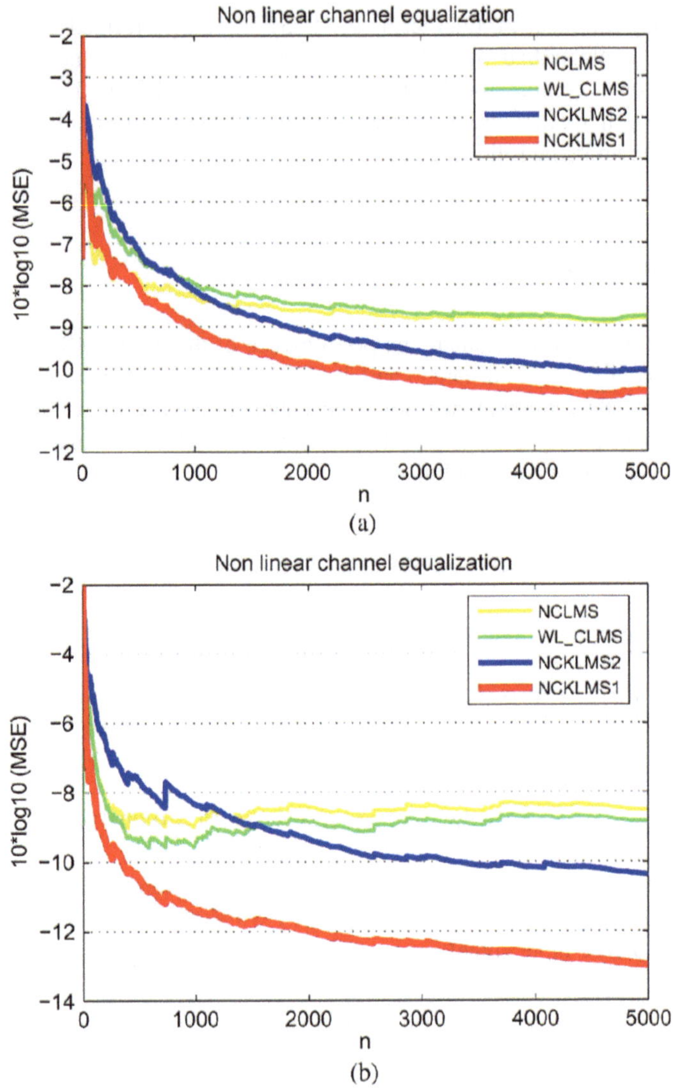


**Fig. 4.1** Learning curves for CKLMS1 ($\mu = 1/2$), CKLMS2 ($\mu = 1/4$), CLMS ($\mu = 1/16$) and WL-CLMS ($\mu = 1/16$) (filter length $L = 5$, delay $D = 2$) for the soft nonlinear channel equalization problem, for **a** the circular input case, **b** the noncircular input case ($\rho = 0.1$). (Reproduced from [1])

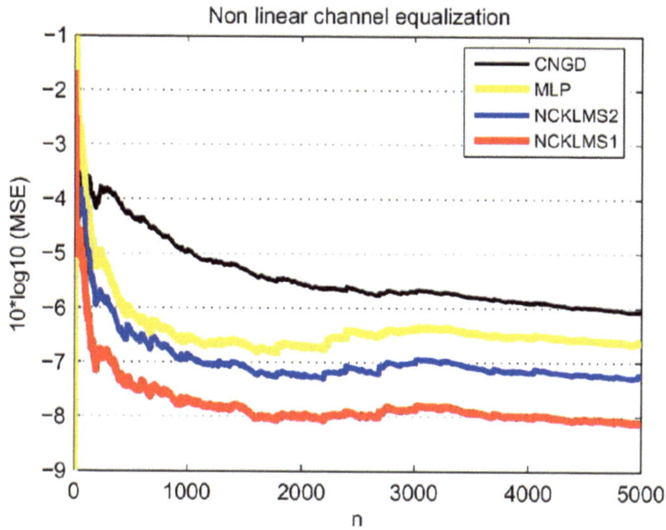


(a)

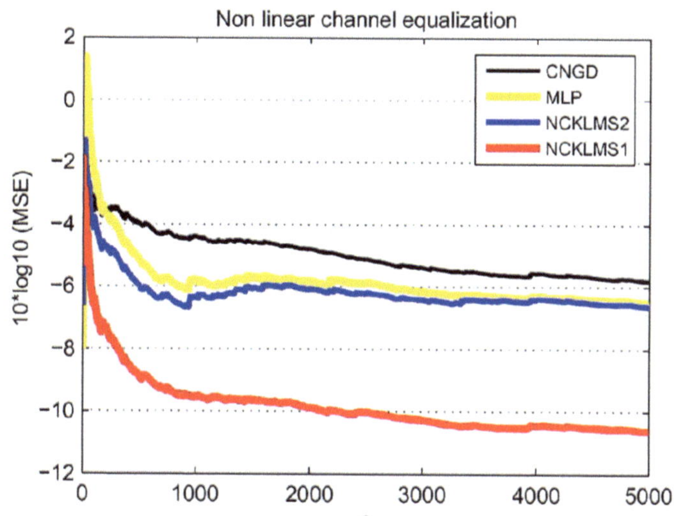


(b)

**Fig. 4.2** Learning curves for CKLMS1 ($\mu = 1/2$), CKLMS2 ($\mu = 1/4$), CNGD and $L$-50–1 MLP (filter length $L = 5$, delay $D = 2$) for the hard nonlinear channel equalization problem, for **a** the circular input case, **b** the noncircular input case ($\rho = 0.1$). (Reproduced from [1])

The NCKLMS-1 algorithm used a "complexified" RKHS formed based on a real kernel (see Sect. 4.1). For NCKLMS-2, a pure complex kernel was used.

Specifically, the kernel functions used were.

Real Gaussian Kernel: $\kappa(x, y) = exp(-\sum_{i=1}^{N}(x_i - y_i)^2/\sigma^2)$.

Complex Gaussian Kernel: $\kappa(z, w) = exp(-\sum_{i=1}^{N}(z_i - w_i^*)^2/\sigma^2)$.

Note the conjugation of $w_i$ in the complex form. For the kernel definitions $x, y \in R^N$ , $z, w \in C^N$ and $\sigma$ is the kernel parameter affecting the transform mapping ($\sigma =$ 5 was used for both kernel functions, and all tests shown).

For Fig. 4.1, the NCKLMS-1, NCKLMS-2 algorithms were compared with the Normalized CLMS (NCLMS) and Widely Linear CLMS (WL-CLMS) algorithms. The new kernel-based methods provided a reduced MSE by more than 1dB for both circular and noncircular data versus the existing linear methods for the soft nonlinear channel.

For Fig. 4.2, NCKLMS-1, NCKLMS-2 were compared with the CNGD and multilayer perceptron (MLP) algorithms [1]. The NCKLMS-1 algorithm provided a significantly reduced MSE (more than 4dB), particularly for noncircular data for the strong nonlinear channel.

The NCKLMS-2 performance for both Figs. 4.1 and 4.2 improved upon existing methods, however was substantially worse than NCKLMS-1. This indicates a nonequivalence between NCKLMS-1 and NCKLMS-2, which is explored further in Chaps. 8 and 9.

Some additional testing and analysis showing the trade off between MSE and computational complexity for CKLMS versus Complex LMS is shown in Appendix B.

## References


1. P. Bouboulis, S. Theodoridis, "Extension of Wirtinger's calculus to reproducing Kernel Hilbert spaces and the complex kernel LMS". IEEE Trans. Signal Process., **59**(3), (2011)
2. P. Bouboulis, S. Theodoridis, "The complex Gaussian kernel LMS algorithm". in *Proc of int conf on artificial neural networks* (2010)
3. D. Westwick, R. Kearney, *Identification of nonlinear physiological systems* (IEEE Press, John Wiley InterScience, 2003)
4. T. Ogunfunmi, *Adaptive nonlinear system identification: The Volterra and Wiener model approaches*, published by Springer Publishers (2007)

# Chapter 5
# Learning with Quaternions

In this chapter, we provide a description of the structure and use of quaternions for adaptive filtering. An introduction to quaternion-valued data is initially outlined, describing some background, as well as the benefits of their use. Afterwards, a calculus for finding gradients of quaternion-based functions, referred to as $\mathbb{HR}$ calculus, is described. Finally, the design of quaternion adaptive filters is considered, which also includes a discussion of widely linear filtering using quaternion-valued data.

## 5.1 Background on Quaternions

In this section, we provide a brief introduction to quaternions. This includes reasons for their use, as well as quaternion definitions, operations, and properties. We also describe some characteristics of the probability distributions of quaternions before their application in later sections.

### *5.1.1 Why Use Quaternions?*

The use of quaternion-valued data has been drawing recent interest in various areas of statistical signal processing, including adaptive filtering, image pattern recognition, and modeling and tracking of motion [1–6]. In particular, linear and widely linear adaptive filter designs using quaternion data are described in references [1, 2]. The $\mathbb{HR}$ calculus for performing gradient operations (described in Section 5.2) was introduced in [3]. Additional examples are described in [4–9].

A particular benefit for quaternion-valued processing involves performing data transformations in 3 or 4-dimensional space in a convenient fashion compared to



T. K. Paul and T. Ogunfunmi, *Adaptive Filter Algorithms For Complex and Quaternion Data*, https://doi.org/10.1007/978-3-032-23798-9_5

vector algebra. These transformations are performed using quaternion addition and multiplication, which differ from real and complex-valued multiplication in that quaternion multiplication is non-commutative. The non-commutative multiplication for quaternions has properties similar to the vector (or cross) product, which is illustrated in the next subsection. The non-commutativity allows 3 or 4-dimensional transformations to be performed conveniently with quaternions, similar to 2-dimensional transformations with complex data.

### *5.1.2 Definitions and Properties*

Quaternions may be described as an associative but not commutative algebra on a real field $\mathbb{R}$, defined as [3].

$$\mathbb{H} = \{q_a + iq_b + jq_c + kq_d | q_a, q_b, q_c, q_d \in \mathbb{R}\}$$

where *i*, *j*, *k* are imaginary units (also considered axis vectors), with $i^2 = j^2 = k^2 = ijk = -1$. Any quaternion value can be expressed as

$$q = q_a + iq_b + jq_c + kq_d = Sq + Vq \tag{5.1}$$

where $\mathrm{R}(q) = Sq = q_a$ is the scalar (real) part, and the vector (or pure quaternion) part is $\mathrm{J}(q) = Vq = iq_b + jq_c + kq_d$. The vector part is formed from the three imaginary parts of quaternions.

The product of two quaternions $q_1, q_2 \in \mathbb{H}$ is

$$q_1q_2 = Sq_1Sq_2 - Vq_1 \cdot Vq_2 + Sq_2Vq_1 + Sq_1Vq_2 + Vq_1 \times Vq_2 \tag{5.2}$$

where symbol "·" is scalar product, and "×" is vector product. Since the definition contains vector product, therefore the product is noncommutative: $q_1q_2 \neq q_2q_1$. Additionally, $ij = -ji = k$, $jk = -kj = i$, and $ki = -ik = j$.

The scalar product $q_1 \cdot q_2 = \langle q_1, q_2 \rangle$ is defined.

$$q_1 \cdot q_2 = q_{1,a}q_{2,a} + q_{1,b}q_{2,b} + q_{1,c}q_{2,c} + q_{1,d}q_{2,d} = \Re\left(q_1q_2^*\right) = \Re\left(q_1^*q_2\right) \tag{5.3}$$

Some additional properties are

$$\begin{aligned} q^* &= Sq - Vq \\ ||q|| &= \sqrt{\langle q, q \rangle} = \sqrt{qq^*} \\ q^{-1} &= q^*/q^2 \end{aligned}$$

and $i^{-1} = -i$, $j^{-1} = -j$, $k^{-1} = -k$.

### 5.1.3 Quaternion Involutions

A similarity relation between two quaternions $q_1$ and $q_2$ (denoted $q_1 \sim q_2$) is said to exist if there exists $\mu \in \mathbb{H}$ such that $q_1 = \mu q_2 \mu^{-1}$, i.e.

$$q_1 \sim q_2 \Leftrightarrow q_1 = \mu q_2 \mu^{-1}, \mu \neq 0 \tag{5.4}$$

This is an equivalence relation, and has the properties that $\|q_1\| = \|q_2\|$ and $q_{1,a} = q_{2,a}$. Note that for quaternions $i \sim j \sim k$ and $q \sim q^*$.

It has been shown that there exist the following equivalence relations for an arbitrary quaternion, $q$ [3]:

$$\begin{aligned} q^i &= -iqi = q_a + iq_b - jq_c - kq_d \\ q^j &= -jqj = q_a - iq_b + jq_c - kq_d \\ q^k &= -kqk = q_a - iq_b - jq_c + kq_d \end{aligned} \tag{5.5}$$

These relations are called quaternion involutions (self-inverse mappings). Note that conjugation can also be considered an involution, with $(q^*)^* = q$.

Based on the involutions of (5.5), each of the real components of a quaternion value can be expressed using the following relations.

$$\begin{aligned} q_a &= \frac{1}{4}\left[q + q^i + q^j + q^k\right], q_b = \frac{-i}{4}\left[q + q^i - q^j - q^k\right] \\ q_c &= \frac{-j}{4}\left[q - q^i + q^j - q^k\right], q_d = \frac{-k}{4}\left[q - q^i - q^j + q^k\right] \end{aligned} \tag{5.6}$$

The involutions were used in [3] for obtaining the $\mathbb{HR}$ calculus.

### 5.1.4 Probability Distributions

For understanding the probability distribution of quaternion-valued signals, the key characteristic is properness, which may be referred to as the dependence of the probability distribution on angle rotations. A quaternion signal is called $\mathbb{Q}$-proper if its pdf is invariant to arbitrary phase rotations w.r.t any of the following six axis pairs [1]:

$$\{1, i\}, \{1, j\}, \{1, k\}, \{i, j\}, \{j, k\}, \{k, i\}$$

Otherwise, the signal is considered $\mathbb{Q}$ -improper (see [1–3]).

A quaternion signal having the characteristic that its real-valued components, i.e., components in 1, $i$, $j$, and $k$ directions, all have identical distributions leads to a $\mathbb{Q}$ -proper signal (invariance under rotation), while non-identical distributions lead to a $\mathbb{Q}$ -improper signal. Properness can be seen to affect the convergence behavior of quaternion adaptive algorithms, which will be visible later in the chapter.

## 5.2 HR Calculus

In a similar way to the Wirtinger's (or $\mathbb{CR}$) calculus, the $\mathbb{HR}$ calculus is obtained by considering a duality between the derivatives of a quaternion function $f(q) \in \mathbb{H}$, and the derivatives of a corresponding real function $g(q_a, q_b, q_c, q_d) \in \mathbb{R}^4$.

Considering that $\mathbb{H}$ and $\mathbb{R}^4$ are isomorphic.

$$\begin{aligned} f(q) &= f_a(q_a, q_b, q_c, q_d) + if_b(q_a, q_b, q_c, q_d) \\ &\quad + jf_c(q_a, q_b, q_c, q_d) + kf_d(q_a, q_b, q_c, q_d) \\ &= g(q_a, q_b, q_c, q_d) \end{aligned} \tag{5.7}$$

Finding the differential of (5.7) gives us

$$\begin{aligned} dg &= \frac{\partial g}{\partial q_a}dq_a + \frac{\partial g}{\partial q_b}dq_b + \frac{\partial g}{\partial q_c}dq_c + \frac{\partial g}{\partial q_d}dq_d \\ &= \frac{\partial f(q)}{\partial q_a}dq_a + i\frac{\partial f(q)}{\partial q_b}dq_b + j\frac{\partial f(q)}{\partial q_c}dq_c + k\frac{\partial f(q)}{\partial q_d}dq_d \end{aligned} \tag{5.8}$$

From the properties (5.6), we can also write

$$\begin{aligned} dq_a &= \frac{1}{4}\left[dq + dq^i + dq^j + dq^k\right], dq_b = \frac{-i}{4}\left[dq + dq^i - dq^j - dq^k\right] \\ dq_c &= \frac{-j}{4}\left[dq - dq^i + dq^j - dq^k\right], dq_d = \frac{-k}{4}\left[dq - dq^i - dq^j + dq^k\right] \end{aligned} \tag{5.9}$$

Since the components of $q$ can be expressed in terms of the perpendicular involutions $q^i, q^j, q^k$ [3], we can write

$$f(q) = f(q_a, q_b, q_c, q_d) = f\left(q, q^i, q^j, q^k\right). \tag{5.10}$$

In terms of the involutions, we can express the differential as

$$\begin{aligned} df(q) = &\frac{\partial f\left(q, q^i, q^j, q^k\right)}{\partial q}dq + \frac{\partial f\left(q, q^i, q^j, q^k\right)}{\partial q^i}dq^i \\ &+ \frac{\partial f\left(q, q^i, q^j, q^k\right)}{\partial q^j}dq^j + \frac{\partial f\left(q, q^i, q^j, q^k\right)}{\partial q^k}dq^k \end{aligned} \tag{5.11}$$

Finally, if the real quadrivariate function $g(q_a, q_b, q_c, q_d) \in \mathbb{R}^4$ is differentiable with respect to real variables $q_a, q_b, q_c, q_d$, we can form the set of $\mathbb{HR}$ derivatives (based on considering Eqs. (5.8)–(5.11)), which are

$$\begin{bmatrix} \partial f\left(q, q^i, q^j, q^k\right)/\partial q \\ \partial f\left(q, q^i, q^j, q^k\right)/\partial q^i \\ \partial f\left(q, q^i, q^j, q^k\right)/\partial q^j \\ \partial f\left(q, q^i, q^j, q^k\right)/\partial q^k \end{bmatrix} = \frac{1}{4}\begin{bmatrix} 1 & -i & -j & -k \\ 1 & -i & j & k \\ 1 & i & -j & k \\ 1 & i & j & -k \end{bmatrix}\begin{bmatrix} \partial f/\partial q_a \\ \partial f/\partial q_b \\ \partial f/\partial q_c \\ \partial f/\partial q_d \end{bmatrix} \tag{5.12}$$

where f is short for $f(q_a, q_b, q_c, q_d)$.

The conjugate derivative form (referred to as the $\mathbb{HR}^*$ derivatives) can be derived based on quaternion conjugate $q^*$, and conjugate involutions $q^{i^*}$, $q^{j^*}$, $q^{k^*}$. The set $\{q^*, q^{i^*}, q^{j^*}, q^{k^*}\}$ can also be seen to form an alternative basis, i.e.

$$f(q) = f(q_a, q_b, q_c, q_d) = f\left(q^*, q^{i*}, q^{j*}, q^{k*}\right).$$

In this case, we can write

$$\begin{aligned} dq_a &= \frac{1}{4}\left[dq^* + dq^{i*} + dq^{j*} + dq^{k*}\right], dq_b = \frac{-i}{4}\left[-dq^* - dq^{i*} + dq^{j*} + dq^{k*}\right] \\ dq_c &= \frac{-j}{4}\left[-dq^* + dq^{i*} - dq^{j*} + dq^{k*}\right], dq_d = \frac{-k}{4}\left[-dq^* + dq^{i} + dq^{j*} - dq^{k*}\right] \end{aligned} \tag{5.13}$$

and in terms of the conjugate involutions

$$\begin{aligned} df(q) = {} & \frac{\partial f\left(q^*, q^{i*}, q^{j*}, q^{k*}\right)}{\partial q^*} dq^* + \frac{\partial f\left(q^*, q^{i*}, q^{j*}, q^{k*}\right)}{\partial q^{i*}} dq^{i*} \\ & + \frac{\partial f\left(q^*, q^{i*}, q^{j*}, q^{k*}\right)}{\partial q^{j*}} dq^{j*} + \frac{\partial f\left(q^*, q^{i*}, q^{j*}, q^{k*}\right)}{\partial q^{k*}} dq^{k*} \end{aligned} \tag{5.14}$$

Based on a similar process with the $\mathbb{HR}$ derivatives, the $\mathbb{HR}^*$ derivatives are

$$\begin{bmatrix} \partial f\left(q^*, q^{i*}, q^{j*}, q^{k*}\right)/\partial q^* \\ \partial f\left(q^*, q^{i*}, q^{j*}, q^{k*}\right)/\partial q^{i*} \\ \partial f\left(q^*, q^{i*}, q^{j*}, q^{k*}\right)/\partial q^{j*} \\ \partial f\left(q^*, q^{i*}, q^{j*}, q^{k*}\right)/\partial q^{k*} \end{bmatrix} = \frac{1}{4}\begin{bmatrix} 1 & i & j & k \\ 1 & i & -j & -k \\ 1 & -i & j & -k \\ 1 & -i & -j & k \end{bmatrix}\begin{bmatrix} \partial f/\partial q_a \\ \partial f/\partial q_b \\ \partial f/\partial q_c \\ \partial f/\partial q_d \end{bmatrix} \tag{5.15}$$

The $\mathbb{HR}$ and $\mathbb{HR}^*$ derivatives can be used for performing differentiation with any quaternion function provided that, when expressed as a real quadrivariate function $g(q_a, q_b, q_c, q_d) \in \mathbb{R}^4$, the function is real differentiable as described previously.

Some properties based on the $\mathbb{HR}$ calculus are [3].

1. If $f(q) = q$, then $\partial f(q)/\partial q = 1$ and $\partial f(q)/\partial q^\eta = 0$ for $\eta \in \{i, j, k\}$.
2. If $f(q) = q^*$, then $\partial f(q)/\partial q = -\frac{1}{2}$ and $\partial f(q)/\partial q^\eta = \frac{1}{2}$ for $\eta \in \{i, j, k\}$ (note this differs from the complex case, where $\partial q^*/\partial q = \partial q/\partial q^* = 0$).

3. If $f(q) = q^*$, then $\partial f(q)/\partial q^* = 1$ and $\partial f(q)/\partial q^{\eta *} = 0$ for $\eta \in \{i, j, k\}$.
4. Derivatives of involutions: $\partial q^\theta/\partial q^\eta = 1$ for $\eta = \theta$ and $\partial q^\theta/\partial q^\eta = 0$ for $\eta \neq \theta$ where $\eta, \theta \in \{i, j, k\}$.

The $\mathbb{HR}$ calculus and properties described may be used for finding the cost function gradients for adaptive filtering algorithms with quaternion data.

### 5.2.1 Involution-Gradient

From [7], the involution-gradient (or i-gradient) was described. The i-gradient is based on an alternative method for defining the gradient $\partial f/\partial q^*$. Specifically, the i-gradient views gradient $\partial f/\partial q^*$ in terms of the involution-wise partial derivatives $\partial f/\partial q^i$, $\partial f/\partial q^j$, and $\partial f/\partial q^k$ (instead of using $\partial f/\partial q_a$, $\partial f/\partial q_b$, $\partial f/\partial q_c$, $\partial f/\partial q_d$ as the basis).

Based on the gradients

$$\frac{\partial f\left(q, q^i, q^j, q^k\right)}{\partial q} = \frac{1}{4}\left[\frac{\partial f}{\partial q_a} - i\frac{\partial f}{\partial q_b} - j\frac{\partial f}{\partial q_c} - k\frac{\partial f}{\partial q_d}\right] \tag{5.16}$$

$$\frac{\partial f\left(q, q^i, q^j, q^k\right)}{\partial q^*} = \frac{1}{4}\left[\frac{\partial f}{\partial q_a} + i\frac{\partial f}{\partial q_b} + j\frac{\partial f}{\partial q_c} + k\frac{\partial f}{\partial q_d}\right]$$

it was shown that $\partial f/\partial q^*$ may be expressed, in terms of the involution-wise partial derivatives, as

$$\mathrm{Re}\left[\frac{\partial f}{\partial q^*}\right] = \frac{1}{3}\mathrm{Re}\left[\frac{\partial f}{\partial q^i} + \frac{\partial f}{\partial q^j} + \frac{\partial f}{\partial q^k}\right] \tag{5.17}$$

$$Im\left[\frac{\partial f}{\partial q^*}\right] = Im\left[\frac{\partial f}{\partial q^i} + \frac{\partial f}{\partial q^j} + \frac{\partial f}{\partial q^k}\right]$$

If we define the gradient alternatively as

$$\nabla_{\omega^n} f = \frac{\partial f}{\partial q^i} + \frac{\partial f}{\partial q^j} + \frac{\partial f}{\partial q^k} \tag{5.18}$$

we see that this definition only differs from $\partial f/\partial q^*$ of (5.17) by the magnitude of the real component (i.e. scaled by a factor of 3).

Another interpretation is that the i-gradient approach can be seen to differ from the original gradient definition by ($0.5\ \partial f/\partial q_a$) [7], i.e.

$$\nabla_{\omega^n} f = \frac{\partial f}{\partial q^*} + \frac{1}{2}\frac{\partial f}{\partial q_a}$$

Using the i-gradient can be seen to accelerate the descent for minimizing a real-valued cost function w.r.t. the real component of quaternion $q$ (i.e. $q_a$). This should generally provide faster convergence.

## 5.3 Quaternion Adaptive Filtering

For designing adaptive filtering algorithms using quaternion properties, $\mathbb{HR}$ calculus, the process is essentially similar with the design using real or complex-valued data. For example, we may describe the goal for the algorithm here as finding the quaternion-valued filter coefficients $\mathbf{w}(n)$ which minimizes $|e(n)|^2$, where $e(n) = d(n) - \hat{d}(n)$, based on desired response $d(n)$, output estimate $\hat{d}(n) = \mathbf{w}(n)^T \mathbf{q}(n)$, and input vector $\mathbf{q}(n)$. In this case, $\mathbf{w}(n)$, $\mathbf{q}(n)$, $d(n)$, $\hat{d}(n)$, and $e(n)$ are all quaternion-valued.

Some quaternion filters are the Quaternion LMS (QLMS) and the Widely Linear QLMS (WLQLMS) [3]. The iQuaternion LMS (iQLMS) was also developed in [7] based on the involution gradient. These algorithms are described here.

### 5.3.1 *Quaternion LMS*

For the QLMS algorithm of [3], the cost function specified was $J(n) = |e(n)|^2 = e(n)e^*(n)$, and the filter output $\hat{d}(n) = \mathbf{w}(n)^T \mathbf{q}(n)$. The weight update for LMS may be written as

$$\boldsymbol{w}(n) = \boldsymbol{w}(n-1) - \mu \nabla_{\boldsymbol{w}*} J(n)$$

Using the $\mathbb{HR}$ calculus, we can write the gradient $\nabla_{\boldsymbol{w}*} J(n)$ as

$$\nabla_{\boldsymbol{w}*} J(n) = e(n) \frac{\partial e^*(n)}{\partial \boldsymbol{w}^*(n)} + \frac{\partial e(n)}{\partial \boldsymbol{w}^*(n)} e^*(n) \tag{5.19}$$

Since the error is: $e(n) = d(n) - \mathbf{w}(n)^T \mathbf{q}(n)$, therefore the gradients of $e(n)$, $e^*(n)$ are

$$\frac{\partial e^*(n)}{\partial \boldsymbol{w}^*(n)} = -\boldsymbol{q}^*(n), \ \frac{\partial e(n)}{\partial \boldsymbol{w}^*(n)} = \frac{1}{2} \boldsymbol{q}(n)$$

and the weight update is

$$\boldsymbol{w}(n) = \boldsymbol{w}(n-1) + \mu \left( e(n) \boldsymbol{q}^*(n) - \frac{1}{2} \boldsymbol{q}(n) e^*(n) \right) \tag{5.20}$$

### 5.3.2 iQuaternion LMS

For the iQLMS algorithm, the problem formulation is similar to QLMS. With cost function $J(n) = e(n)e^*(n)$, and output $\hat{d}(n) = \mathbf{w}(n)^H \mathbf{q}(n)$, we again use

$$\boldsymbol{w}(n) = \boldsymbol{w}(n-1) - \mu \nabla_{\boldsymbol{w}*} J(n)$$

However, here we evaluate $\nabla_{\boldsymbol{w}*} J(n)$, i.e.:

$$\nabla_{\boldsymbol{w}*} J(n) = e(n) \frac{\partial e^*(n)}{\partial \boldsymbol{w}^*(n)} + \frac{\partial e(n)}{\partial \boldsymbol{w}^*(n)} e^*(n)$$

based on the i-gradient, using the property

$$\boldsymbol{w}^* = \frac{1}{2}\left(\boldsymbol{w}^i + \boldsymbol{w}^j + \boldsymbol{w}^k - \boldsymbol{w}\right)$$

For both $e(n)$ and $e^*(n)$, it can be seen that

$$\frac{\partial e(n)}{\partial \boldsymbol{w}^\beta} = -\frac{1}{2} \boldsymbol{q}(n), \ \frac{\partial e^*(n)}{\partial \boldsymbol{w}^\beta} = 0 \quad \forall \beta \in \{i, j, k\} \tag{5.21}$$

and based on Eq. (5.18), we get

$$\nabla_{\boldsymbol{w}*}\{|e(n)|^2\} = -\frac{3}{2} \boldsymbol{q}(n) e^*(n)$$

Therefore the iQLMS weight update is

$$\boldsymbol{w}(n) = \boldsymbol{w}(n-1) + \mu \boldsymbol{q}(n) e^*(n) \tag{5.22}$$

where the scaling is included in $\mu$. The update form is seen to match Complex LMS with this approach [7].

### 5.3.3 Quaternion Widely Linear Model

Similar to the complex data case, the estimation form $\hat{d}(n) = \mathbf{w}(n)^T \mathbf{q}(n)$ may not be sufficient to represent the input statistics for mean square estimation regression with quaternions. An approach similar to the one proposed for complex data (i.e. widely linear estimation) is described in [2, 3], and summarized here.

With the quaternion widely linear (QWL) model, the estimation problem is viewed in the following way. Initially, consider the real-valued minimum mean square error

(MMSE) estimator case. Here, the estimator is: $\hat{y} = E[y|x]$, where $\hat{y}$ is the MMSE estimate of signal $y$, given observation $x$.

If $y$ and $x$ are zero mean, jointly normal, then the solution is linear, i.e.

$$\hat{y} = E[y|x] \rightarrow y = \boldsymbol{w}^T \boldsymbol{x}$$

where $\boldsymbol{x} = [x_1 \ldots x_N]^T$ is a vector of past observations, and $\mathbf{w}$ is a coefficient vector.

Similarly, for the quaternion case, the standard conditional MSE estimate would be $\hat{y} = E[y|q], \forall q, y \in \mathbb{H}$. However, note that in terms of the real components, we can write

$$\hat{y}_{\{m\}} = E[y_{\{m\}}|q_a, q_b, q_c, q_d], m \in \{a, b, c, d\}$$

Since the real-valued components $q_a, q_b, q_c, q_d$ can be fully expressed in terms of the quaternion perpendicular involutions, we can write

$$\hat{y} = E[y|q, q^i, q^j, q^k]$$

This leads to the quaternion widely linear (QWL) model

$$\hat{y} = \boldsymbol{u}^T \boldsymbol{q} + \boldsymbol{v}^T \boldsymbol{q}^i + \boldsymbol{g}^T \boldsymbol{q}^j + \boldsymbol{h}^T \boldsymbol{q}^k = \boldsymbol{w}^{aT} \boldsymbol{x}^a \tag{5.23}$$

where $\mathbf{u}$, $\mathbf{v}$, $\mathbf{g}$, $\mathbf{h}$ are quaternion coefficient vectors, $\boldsymbol{q} = [q_1 \ldots q_N]^T$, $\boldsymbol{w}^a = [\boldsymbol{u}^T \boldsymbol{v}^T \boldsymbol{g}^T \boldsymbol{h}^T]^T$ and $\boldsymbol{x}^a = [\boldsymbol{q}^T \boldsymbol{q}^{iT} \boldsymbol{q}^{jT} \boldsymbol{q}^{kT}]^T$ (referred to as the augmented input). Equation (5.23) is the estimator considered for the algorithm in the next section.

Note that the use of the QWL model is beneficial when handling inputs of unknown distribution, in particular the notion of properness of quaternion random variables (i.e. $\mathbb{Q}$-proper and $\mathbb{Q}$-improper). Widely linear estimation allows for optimal minimum MSE (mean-square-error) modeling for both types of signals [2, 3].

### 5.3.4 Widely Linear Quaternion LMS

For the Widely Linear Quaternion LMS (WLQLMS) algorithm, the cost function remains $J(n) = |e(n)|^2$, however the filter output has the form $\hat{d}(n) = \mathbf{w}^a(n)^T \mathbf{x}^a(n)$, where $\mathbf{w}^a$(n) are the quaternion weights, and $\boldsymbol{x}^a(n) = [\boldsymbol{q}^T(n)\boldsymbol{q}^{iT}(n)\boldsymbol{q}^{jT}(n)\boldsymbol{q}^{kT}(n)]^T$ the augmented input. The weight update can again be described by

$$\boldsymbol{w}(n) = \boldsymbol{w}(n-1) - \mu \nabla_{\boldsymbol{w}*} J(n)$$

Gradient $\nabla_{w*} J(n)$, using $\mathbb{HR}$ calculus, can be shown to be (similar to before):

$$\nabla_{w*} J(n) = e(n) \frac{\partial e^*(n)}{\partial \boldsymbol{w}^*(n)} + \frac{\partial e(n)}{\partial \boldsymbol{w}^*(n)} e^*(n) \tag{5.24}$$

where

$$\frac{\partial e^*(n)}{\partial \boldsymbol{w}^*(n)} = -\boldsymbol{x}^{a*}(n), \frac{\partial e(n)}{\partial \boldsymbol{w}^*(n)} = \frac{1}{2}\boldsymbol{x}^a(n)$$

Finally, the weight update for WLQLMS is

$$\boldsymbol{w}(n) = \boldsymbol{w}(n-1) + \mu\left(e(n)\boldsymbol{x}^{a*}(n) - \frac{1}{2}\boldsymbol{x}^a(n)e^*(n)\right) \tag{5.25}$$

Simulation results for the WLQLMS and QLMS adaptive algorithms are shown in Sect. 5.4.

## 5.4 Simulations

The WLQLMS and QLMS algorithms were compared with simulations in [2], and the results obtained are reproduced here. The simulation test results included here are for a $\mathbb{Q}$-proper synthetic AR(4) process, a $\mathbb{Q}$-improper 4-dimensional (4D) Saito's signal, as well as a real-world 4D wind field signal.

The results from Fig. 5.1 show the performance of the WLQLMS, QLMS, and AQLMS (Augmented QLMS) algorithms for the 4D Saito's signal. The AQLMS is an earlier method shown to use incomplete statistics versus WLQLMS [2], and is not considered here. From the simulations, WLQLMS is shown to yield better convergence behavior, as well as MSE performance, compared with QLMS.

Similarly, the performance of the algorithms against the 4D wind model is shown in Fig. 5.2. In this test case, the WLQLMS convergence performance also exceeds that of QLMS. The MSE performance in this case is similar, however. For both Figs. 5.1 and 5.2, the testing data used is $\mathbb{Q}$-improper, and benefit may be seen using WLQLMS with these cases.

Finally for Fig. 5.3, the prediction performance of the algorithms for a $\mathbb{Q}$-proper AR(4) process is shown. The AR process used was

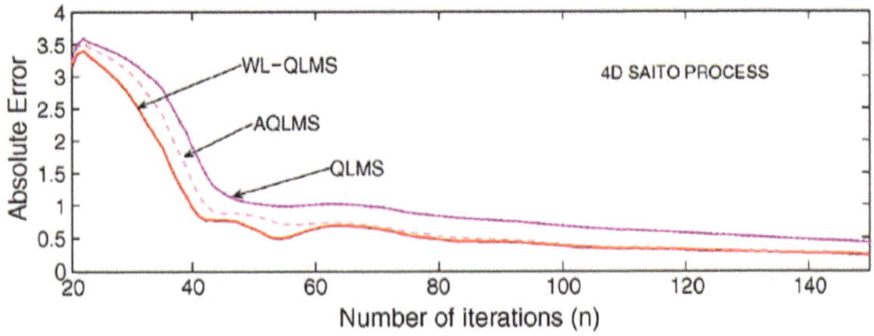


**Fig. 5.1** WLQLMS, QLMS for 4D Saito process (from [2])

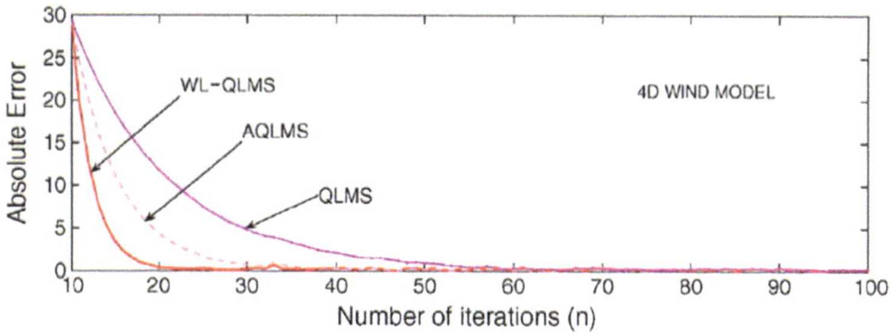


**Fig. 5.2** WLQLMS, QLMS for 4D wind model (from [2])

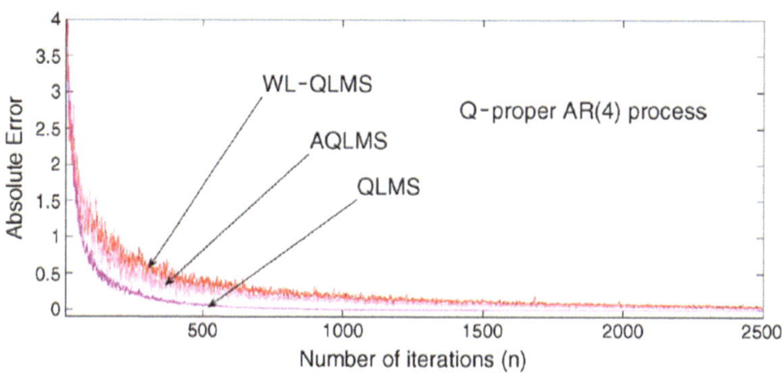


**Fig. 5.3** WLQLMS, QLMS for $\mathbb{Q}$-proper AR(4) process (from [2])

$$x(n) = 1.79x(n-1) - 1.85x(n-2) + 1.27x(n-3) - 0.41x(n-4)$$

and was driven by quadruply white Gaussian noise, with iid real, imaginary components. In this case, the QLMS provided faster convergence and good MSE compared to WLQLMS. Note with $\mathbb{Q}$-proper data, the benefit of WLQLMS was not substantial, and the additional dimension size of this filter can be seen to reduce performance.

The quaternion-based adaptive algorithms described here are extended for kernel filtering in later chapters.

## References


1. S. Javidi, C. Cheong Took, C. Jahanchahi, N. Le Bihan, D.P. Mandic, Blind Extraction of improper quaternion sources. in *IEEE Proceedings of Inter. Conf. on Acoustics, Speech and Signal Processing (ICASSP)* (2011), pp. 3078–3711. http://www.commsp.ee.ic.ac.uk/~mandic/research
2. C. Cheong Took, D.P. Mandic, A quaternion widely linear adaptive filter. IEEE Trans. Signal Process., **58**(8), 4427–4431 (2010)

3. D.P. Mandic, C. Jahanchahi, C. Cheong Took, A quaternion gradient operator and its applications. IEEE Signal Process. Lett., **18**(1), 47–50 (2011)
4. A. Shilton, D.T.H. Lai, Quaternionic and complex-valued support vector regression for equalization and function approximation. in *Proc of inter joint conf on neural networks* (2007)
5. E.J.B.-Corrochano, N.A.-Daniel, Clifford support vector machines for classification, regression, and recurrence. IEEE Trans. Neural Netw., **21**(11), 1731–1746 (2010)
6. T. Shaneyfelt, S. Agaian, M. Jamshidi, S. Erdogan, Quaternion number based vanilla recognition system for automating polination. in *Proc of inter conf. on system science and engineering* (2011), pp. 558–563
7. C. Jahanchahi, C. Cheong Took, D. P. Mandic, On gradient calculations in quaternion adaptive filtering. in *Proc of Inter Conf on Acoustics, Speech, and Signal Processing (ICASSP)* (2012), pp. 3773–3776
8. A. Shilton, Mercer's theorem for quaternionic kernels, Univ. Melbourne, Dept. Elect. Electron. Eng., Melbourne, Australia, Tech. Rep. TR07–009. [Online]. http://www.ee.unimelb.edu.au/staff/apsh/publications/
9. A. Sudberry, Quaternionic analysis. Math. Proc. Cambridge Philos. Soc. **85**, 199–225 (1979)

# Chapter 6
# Information Theoretic Learning

Here we provide a brief introduction to information theory, as well as the area of information-theoretic learning (ITL). Use of ITL has had a growing interest, because it enables the design of adaptive nonlinear filters that make limited assumptions on the statistics of signals in the system (for example, signal distributions need not be assumed Gaussian). ITL focuses instead on information content for processing signals.

This chapter initially provides a brief background on information theory and outlines some of the common information theoretic measures used. Afterwards, the use of a new measure, correntropy, is described for improving adaptive filter performance with non-Gaussian signals compared to mean square error (MSE). This correntropy-based approach is used for deriving some of the algorithms in later chapters.

## 6.1 Background on Information Theory

The origins of information theory start from the 1940's. Claude Shannon provided proof that the probability of error for transmitting across a communication channel can be made arbitrarily close to zero for any rate of transmission below the channel capacity. In order to show this, he used an approach that forms the basis of information theory.

He found that the capacity for channels may be determined from the characteristics of the noise present. Also, the random processes underlying any transmission (for example, music or speech) can be seen to have an inherently fundamental complexity, below which the transmission cannot be compressed. Shannon referred to this as entropy, noting its similarity to this term from thermodynamics. If the entropy of a signal is less than the capacity of a channel, he argued that asymptotically error-free communication may be performed.



T. K. Paul and T. Ogunfunmi, *Adaptive Filter Algorithms For Complex and Quaternion Data*, https://doi.org/10.1007/978-3-032-23798-9_6

Information theory has evolved to represent information in various fields (i.e. physics, computer science, economics, mathematics, probability, statistics) along with communications. In communications, it allows for describing the performance of any possible method of data transfer. The minimum required rate, based on maximum data compression, represents one side of the possible rates, while the maximum rate is bounded by channel capacity. Therefore, information theory indicates the limits on the ability to transfer any source data across a particular channel.

Note that this section was meant to provide a basic conceptual background on areas of information theory only, and the interested reader can refer to references such as [1–4] for theoretical foundations.

Information theory can be used to improve the optimization goal for adaptive filtering. The use of information theory in this direction is referred to as Information Theoretic Learning (ITL), which is discussed in this chapter.

The traditional approach for adaptive filtering typically involves the use of second-order signal statistics (described in Chap. 2). This is mainly motivated by reducing algorithm complexity, as well as ensuring the existence of analytical solutions. Current improvements in computational power, and modern signal processing challenges where existing methods may be inadequate have generated interest in ITL methods. Applications for use of ITL include system identification, blind source separation, data clustering, image registration, etc. [5].

The ITL approaches are based on measuring, analyzing information content for performing nonlinear adaptive filtering tasks. Information measures may be used to determine the amount of uncertainty in a signal (i.e. entropy), or compare the information contained in two signals (i.e. divergence, mutual information). They can be obtained via nonparametric sample estimators and serve as a basis for determining the accuracy of an adaptive filter in a more precise way than second-order statistics, in general. The information measures described are detailed in the next section.

## 6.2 Information Theoretic Measures

As mentioned earlier, some information measures are entropy, divergence and mutual information. The entropy and divergence measures were proposed in information theory, and are used in determining optimal messages and systems for minimizing the effect of noise on data transmissions [1, 5]. Meanwhile, mutual information is commonly used in communications theory for characterizing statistical dependency between signals. These measures are described here.

Entropy is a measure of uncertainty of some random vector $\mathbf{X}$ possessing a joint probability distribution function (pdf) we refer to as $p(\mathbf{x})$. It is a generalization of variance for processes having non-Gaussian distributions, and was defined by Shannon to be

$$H_S(\mathbf{x}) = -\int p(\mathbf{x}) \log p(\mathbf{x}) d\mathbf{x} = -E\left[\log p(\mathbf{x})\right] \tag{6.1}$$

Divergence is a measure of statistical similarity, and may be viewed as a generalization of algebraic distance to probability distributions. The Kullback–Leibler divergence (KLD) between the distributions $p(\mathbf{x})$ and $q(\mathbf{x})$ is defined.

$$D_{KL}(p; q) = \int p(\boldsymbol{x}) \log \frac{p(\boldsymbol{x})}{q(\boldsymbol{x})} d\boldsymbol{x} \tag{6.2}$$

Note that the divergence measure is asymmetric (i.e. measures the distance of distribution $p$ to $q$). The measure is zero if and only if the distributions are identical; else, the measure is a non-zero positive value.

Mutual information is a special case of KLD, and is a measure of statistical dependence which may be viewed as a generalized form of correlation for arbitrarily nonlinear-transformed multiple processes which have arbitrary distributions. Its definition is

$$I_S(\boldsymbol{x}; \boldsymbol{y}) = \int \int p(\boldsymbol{x}, \boldsymbol{y}) \log \frac{p(\boldsymbol{x}, \boldsymbol{y})}{p(\boldsymbol{x})p(\boldsymbol{y})} d\boldsymbol{x}d\boldsymbol{y} \tag{6.3}$$

These are some of the key measures for ITL, and may be used in deriving algorithms based on information costs instead of second-order (quadratic) costs.

However for computing these measures from sample data, it is important to obtain estimates for the probability distributions of signals in the system. These distributions can be estimated using nonparametric sample estimators. An approach is described next.

### 6.2.1 Parzen Windowing-Based Estimation

Parzen windowing (also referred to as kernel density estimation) is a nonparametric estimation method used for probability distribution as well as entropy estimation [2–5]. The use of Parzen windowing for estimating some distribution $q(x)$ based on input samples is briefly described here.

Given a set of $N$ iid samples $\{x_1, x_2, \ldots x_N\}$ obtained from distribution $q(x)$, the distribution estimate based on Parzen windowing is

$$p_N(x) = \frac{1}{N} \sum_{i=1}^{N} K_\sigma(x - x_i) \tag{6.4}$$

where $K_\sigma(\xi)$ is assumed a fixed-size kernel function, for simplicity. The kernel and size $\sigma$ can be optimized using the maximum likelihood (ML) principle, or other methods [5].

For a particular kernel function, the Parzen window estimate has the following properties:

1. For fixed $\sigma$, $E\left[p_N(x)\right] = \lim_{N\to\infty}, p_N(x) = q(x)K_\sigma(x)$
2. For fixed $\sigma$, $\lim_{N\to\infty}, \operatorname{var}\left[p_N(x)\right] = 0$
3. If $\lim_{N\to\infty}, \sigma(N) = 0$ and $\lim_{N\to\infty}, N\sigma(N) = \infty$, then $\lim_{N\to\infty}, p_N\ (x) = q(x)$ in probability

These properties should ensure a consistent and asymptotically unbiased estimate for measuring analytic probability distributions.

The use of Parzen windowing in estimating entropy (specifically correntropy) is described in the following section.

## 6.3 Adaptive Filtering with Maximum Correntropy

In this section, we describe an alternative cost function for adaptive filtering based on the maximization of a measure of similarity, known as correntropy, between signals. We initially begin by describing correntropy.

### *6.3.1 Correntropy*

From [6–9], correntropy is a measure that gives a probabilistic estimate of similarity between two random variables. A kernel function $\kappa_\sigma(.)$ is used, where size $\sigma$ adjusts the "window" where similarity is measured.

The definition of correntropy is:

$$V_\sigma(X, Y) = E[\kappa_\sigma(X - Y)] \tag{6.5}$$

where $X$ and $Y$ are random variables (real-valued here), and $\kappa_\sigma(.)$ is a symmetric positive definite kernel. Following [6], for $\kappa_\sigma(.)$ we consider the Gaussian kernel.

With real input data:

$$\kappa_\sigma(X - Y) = \frac{1}{\sqrt{2\pi}\sigma}\exp[\frac{-1}{2\sigma^2}(X - Y)^2] \tag{6.6}$$

For estimating correntropy, the expectation term in (6.5) may be approximated for a finite set of input samples $\{x_1, x_2, \ldots x_N\}$ and $\{y_1, y_2, \ldots y_N\}$ using

$$\hat{V}_{N,\sigma}(X, Y) = \frac{1}{N}\sum_{i=1}^{N}\kappa_\sigma(x_i - y_i) \tag{6.7}$$

Correntropy was shown to have the following properties [6]:

1. The correntropy is positive definite and bounded, i.e.

$$0 \leq \boldsymbol{V}_{\sigma}(\boldsymbol{X}, \boldsymbol{Y}) \leq \frac{1}{\sqrt{2\pi}\boldsymbol{\sigma}}$$

   The maximum value is reached iff $X = Y$.

2. Correntropy is a function of all even moments of term $(X–Y)$, i.e.$V_{\sigma}(X, Y) = \frac{1}{\sqrt{2\pi}\sigma} \sum_{n=0}^{\infty} (-1)^n/(2^n n!) E[(X - Y)^{2n}/(\sigma^{2n})]$

   When $\sigma$ increases, the higher-order terms of $V_{\sigma}(X,Y)$ reduce and the second-order terms dominate. Therefore, the function approximates a quadratic cost function, depending on $\sigma$.

3. With correntropy expressed

$$\boldsymbol{h}_{\sigma}(\boldsymbol{E}) = \boldsymbol{V}_{\sigma}(\boldsymbol{X}, \boldsymbol{Y})$$

   The function $h_{\sigma}(E)$ can be seen as strictly concave in the range $E \in [-\sigma, \sigma]$.

These properties allow for the use of gradient-based adaptive filtering techniques based on the correntropy measure.

### 6.3.2 Maximum Correntropy Criterion

For the use of correntropy in adaptive filters, we may consider the adaptive filter goal of finding an estimate $\hat{d}(n)$ for minimizing error from the desired signal, $d(n)$. Instead of the previous MSE-based cost function (i.e. $E\left[|d(n) - \hat{d}(n)|^2\right]$, approximated for LMS using instantaneous error $\left|d(n) - \hat{d}(n)\right|^2$, here we form the cost function goal as maximizing the correntropy between $\hat{d}(n)$ and $d(n)$.

For achieving this, we can use the cost function:

$$J_n = \frac{1}{N} \sum_{i=n-N+1}^{n} \kappa_{\sigma}(d_i - y_i) \tag{6.8}$$

where $d_i$ is the desired response, $y_i$ is the adaptive filter output and $N$ the number of past samples used for the correntropy estimate.

The objective is maximizing the correntropy (i.e. similarity measure) between $d_i$ and $y_i$, referred to as the Maximum Correntropy Criterion (MCC) in [6].

### 6.3.3 Maximizing Correntropy for Linear Filters

For a linear adaptive filter based on cost function (6.8), the use of gradient ascent may be used (since the goal is maximizing $J_n$ here).

With linear estimator $y_n = \mathbf{w}_n{}^T \mathbf{u}_n$, where $\mathbf{u}_n$ is the system input and $\mathbf{w}_n$ is the weight vector, the update for vector $\mathbf{w}_n$ based on gradient ascent is

$$\begin{aligned} \boldsymbol{w}_{n+1} &= \boldsymbol{w}_n + \eta \frac{\partial}{\partial \boldsymbol{w}_n^*}[J_n] \\ &= \boldsymbol{w}_n + \eta \frac{\partial}{\partial \boldsymbol{w}_n^*}\left[\kappa_\sigma(d_n - y_n)\right] \\ &= \boldsymbol{w}_n + \eta \frac{\partial}{\partial \boldsymbol{w}_n^*}\{\exp[\frac{-(d_n - \boldsymbol{w}_n^T \boldsymbol{u}_n)^2}{2\sigma^2}]\} \\ &= \boldsymbol{w}_n + \eta \exp[\frac{-(d_n - \boldsymbol{w}_n^T \boldsymbol{u}_n)^2}{2\sigma^2}]\left(\frac{2(d_n - \boldsymbol{w}_n^T \boldsymbol{u}_n)\boldsymbol{u}_n}{2\sigma^2}\right) \\ &= \boldsymbol{w}_n + \eta \exp[\frac{-e_n^2}{2\sigma^2}]e_n \boldsymbol{u}_n \end{aligned} \tag{6.9}$$

where $e_n = d_n - \mathbf{w}_n{}^T \mathbf{u}_n$, and where any scaling terms are included in step-size, $\eta$. Note that $N = 1$ is assumed for (6.8), i.e. an instantaneous correntropy estimate.

This approach, referred to as MCC, has similar complexity to the LMS algorithm. From the properties described earlier, with a suitable choice of $\sigma$, the cost function is sufficiently smooth and should be a statistically robust method. The benefits of the use of the maximum correntropy criterion cost function for modeling non-Gaussian signals can be seen later in the chapter.

### 6.3.4 Maximizing Correntropy for Kernel Filters

The derivation of a kernel adaptive filter based on cost function (6.8) is similar to the design for linear filters, aside from the kernel-based transformation of the input. The gradient ascent will again be used for maximizing correntropy.

The estimator now has form: $y_n = \mathbf{w}_n{}^T \boldsymbol{\varphi}_n$, where $\boldsymbol{\varphi}_n$ is shorthand for kernel transformation $\varphi(\mathbf{u}_n)$ of the system input, and $\mathbf{w}_n$ is the weight vector.

The update for vector $\mathbf{w}_n$ based on gradient ascent (again assuming $N = 1$):

$$\begin{aligned} \boldsymbol{w}_{n+1} &= \boldsymbol{w}_n + \eta \frac{\partial}{\partial \boldsymbol{w}_n^*}[J_n] \\ &= \boldsymbol{w}_n + \eta \frac{\partial}{\partial \boldsymbol{w}_n^*}\left[\kappa_\sigma(d_n, y_n)\right] \end{aligned}$$

$$= \boldsymbol{w}_n + \eta \frac{\partial}{\partial \boldsymbol{w}_n^*} \{\exp[\frac{-(\boldsymbol{d}_n - \boldsymbol{w}_n^T \boldsymbol{\varphi}_n)^2}{2\sigma^2}]\}$$

$$= \boldsymbol{w}_n + \eta \exp[\frac{-e_n^2}{2\sigma^2}] e_n \varphi_n \tag{6.10}$$

where $e_n = d_n$ - $\mathbf{w}_n{}^T \varphi_n$, and scaling terms are again included in step-size, $\eta$.

For the kernel-based method, we need to cast the algorithm in terms of inner products (see Chap. 2). Assuming $\mathbf{w}_0 = 0$, and finding weights at iteration $n$ yields.

$$\begin{aligned}
w_2 &= \eta \exp[\frac{-e_1^2}{2\sigma^2}] e_1 \varphi_1 \\
w_3 &= \eta \exp[\frac{-e_1^2}{2\sigma^2}] e_1 \varphi_1 + \eta \exp[\frac{-e_2^2}{2\sigma^2}] e_2 \varphi_2 \\
&\vdots \\
w_n &= \eta \sum_{i=1}^{n-1} \exp[\frac{-e_i^2}{2\sigma^2}] e_i \varphi_i
\end{aligned} \tag{6.11}$$

And the estimates, expressed in inner products, are

$$\begin{aligned}
y_n = \boldsymbol{w}_n^T \boldsymbol{\varphi}_n &= \left( \eta \sum_{i=1}^{n-1} \exp[\frac{-e_i^2}{2\sigma^2}] e_i \boldsymbol{\varphi}_i \right)^T \boldsymbol{\varphi}_n \\
&= \eta \sum_{i=1}^{n-1} \left( \exp[\frac{-e_i^2}{2\sigma^2}] e_i \boldsymbol{\varphi}_i^T \boldsymbol{\varphi}_n \right) \\
&= \eta \sum_{i=1}^{n-1} \left( \exp[\frac{-e_i^2}{2\sigma^2}] e_i \kappa_\theta(\boldsymbol{u}_i, \boldsymbol{u}_n) \right)
\end{aligned} \tag{6.12}$$

where $\kappa_\theta(.)$ is the kernel function for input transforming. Note $\kappa_\theta(.)$ can differ from $\kappa_\sigma(.)$ used for the correntropy estimate (both in kernel function chosen and size). The resulting algorithm is referred to as the Kernel Maximum Correntropy (KMC) algorithm.

## 6.4 Simulations

Simulations comparing the MCC and KMC algorithms with the KLMS for a nonlinear optimization problem are shown here. The problem considered involves learning of a nonlinear transformation, along with non-Gaussian noise.

The simulation uses an input sine wave at $f_0$, and desired data at $2f_0$, respectively (i.e. frequency doubling). The test uses 1500 samples of these two sequences for training, and 200 samples for test data. An impulsive Gaussian mixture model was used for adding non-Gaussian noise to the system. The probability density function (pdf) of the noise was

$$p_{noise}(i) = 0.9N(0, 0.01) + 0.1N(2.0.01)$$

where $N(m_N, \sigma_N)$ denotes the normal (Gaussian) pdf with mean $m_N$, variance $\sigma_N$.

For comparing KMC and MCC with the KLMS, the following system parameters were used. The input vector used was length 2 (i.e. current and previous input samples). For both kernel methods, the learning rate was 0.9, while for MCC, the rate was 0.2. For the kernel function $\kappa_\theta(.)$ for transforming the input, the kernel size was $\theta = 0.5$ for both KMC and KLMS. The size for $\kappa_\sigma(.)$ of the correntropy estimate was $\sigma = 0.4$ for both KMC and MCC, which was found to perform best on the data. Additionally, 100 Monte Carlo tests were performed.

To guarantee the KMC and MCC algorithms converge towards the global optimum, the algorithms were initially trained using the MSE criterion for the first 300 samples. The results obtained were based on the intrinsic error power (IEP), i.e.

$$IEP = E[d - f(u)]^2$$

where $\boldsymbol{d}$ and $\mathbf{u}$ are the desired and input signal for clean test data, respectively, and $f(\mathbf{u})$ is the system output. The results are shown in Fig. 6.1.

From the graphs, we can see KMC performs better than the other methods, both in terms of the average value of the IEP, as well as the variance. KLMS outperforms

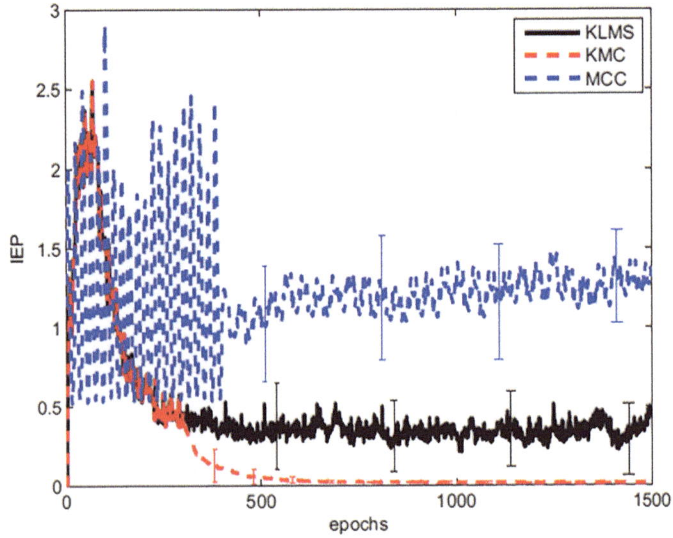


**Fig. 6.1** KMC, MCC, KLMS algorithms (from [6])

MCC for the nonlinear problem; however is outperformed by KMC considering the non-Gaussian signals used.

It should be noted that the choice of kernel parameter $\sigma$ is important to observe this benefit with KMC, and depends on the application. The extension of the KMC algorithm for use with quaternion data is considered in a later chapter.

## References


1. T. Cover, J. Thomas, *Elements of information theory* (Wiley, New York, 1991)
2. J. Beirlant, E.J. Dudewicz, L. Gyorfi, E.C. van der Meulen, Nonparametric entropy estimation: An overview. Int. J. Math. Statist. Sci. **6**(1), 17–39 (1997)
3. J.F. Bercher, C. Vignat, Estimating the entropy of a signal with applications. IEEE Trans. Signal Processing **48**(6), 1687–1694 (2000)
4. E. Parzen, On the estimation of a probability density function and the mode. Ann. Math. Stat. **33**(3), 1065–1076 (1962)
5. D. Erdogmus, J.C. Principe, "From linear adaptive filtering to nonlinear information processing." IEEE Signal Process. Mag., (2006)
6. S. Zhao, B. Chen, J.C. Principe, Kernel adaptive filtering with maximum correntropy criterion. in *Proc of inter joint conf on neural networks* (2011)
7. W. Liu, P. Pokharel, J.C. Principe, Correntropy: Properties and applications in Non-Gaussian signal processing. IEEE Trans. Signal Process. **55**(11), 5286–5298 (2007)
8. W. Liu, P.P. Pokharel, J. C. Principe, "Correntropy: A localized similarity measure," in *Proc Int Joint Conf Neural Networks (IJCNN)* (2006), pp. 4919–4924
9. I. Santamaria, P.P. Pokharel, J.C. Principe, Generalized correlation function: Definition, properties and application to blind equalization. IEEE Trans. Signal Process. **54**(6), 2187–2197 (2006)

# Chapter 7
# The Complex Kernel Affine Projection Algorithm

This chapter presents the development of the Complex Kernel Affine Projection Algorithm (CKAPA).

For the Complex Kernel Least Mean Square (CKLMS) algorithm of [1, 2] (see Chap. 4), rules for a modified Wirtinger's calculus for Hilbert spaces were defined. These allowed the cost gradient for the algorithm, which was defined on a complex RKHS, to be evaluated. Here, the modified calculus was used to obtain the cost function gradient for an APA-based version of the algorithm, which we refer to here as the Complex Kernel Affine Projection Algorithm (CKAPA). Various forms of the algorithm are considered.

## 7.1 Derivation of Algorithm

To derive the complex kernel APA algorithm (CKAPA), we consider the complex APA algorithm derivation of [2], with minor notation differences for consistency with [4]. Keeping the $K$ most recent input vectors and observations:

$$\begin{aligned} \boldsymbol{X}_{ap}(n) &= \boldsymbol{\Phi}\left([z(n-K+1)\ z(n-K+2)\ldots z(n)]^T\right) \\ \boldsymbol{d}_{ap}(n) &= [d(n-K+1)\ d(n-K+2)\ldots\ d(n)]^T \end{aligned} \tag{7.1}$$

The goal of the affine projection algorithm is

$$\begin{aligned} &\text{Minimize}: \frac{1}{2}\|\boldsymbol{w}(n)-\boldsymbol{w}(n-1)\|^2 \\ &\text{Subject to}: \boldsymbol{d}_{ap}(n)-\boldsymbol{X}_{ap}^T(n)\boldsymbol{w}^*(n)=0 \end{aligned} \tag{7.2}$$

The method of Lagrange multipliers is used to convert (7.2) to an unconstrained problem, which becomes [3, 5]



T. K. Paul and T. Ogunfunmi, *Adaptive Filter Algorithms For Complex and Quaternion Data*, https://doi.org/10.1007/978-3-032-23798-9_7

$$F[\boldsymbol{w}(\boldsymbol{n})] = \frac{1}{2}[\boldsymbol{w}(\boldsymbol{n}) - \boldsymbol{w}(\boldsymbol{n}-1)]^{\boldsymbol{H}}[\boldsymbol{w}(\boldsymbol{n}) - \boldsymbol{w}(\boldsymbol{n}-1)] + \mathrm{Re}\left\{\boldsymbol{\lambda}_{ap}^{T}(n)\left[\boldsymbol{d}_{ap}(n) - \boldsymbol{X}_{ap}^{T}(n)\boldsymbol{w}^{*}(n)\right]\right\} \tag{7.3}$$

where $\boldsymbol{\lambda}_{ap}(n)$ is a $K \times 1$ vector of the Lagrange multipliers.

This can be written as

$$\begin{aligned} F[\boldsymbol{w}(\boldsymbol{n})] =& \frac{1}{2}[\boldsymbol{w}(\boldsymbol{n}) - \boldsymbol{w}(\boldsymbol{n}-1)]^{\boldsymbol{H}}[\boldsymbol{w}(\boldsymbol{n}) - \boldsymbol{w}(\boldsymbol{n}-1)] \\ &+ \frac{1}{2}\boldsymbol{\lambda}_{ap}^{H}(\boldsymbol{n})\left[\boldsymbol{d}_{ap}^{*}(\boldsymbol{n}) - \boldsymbol{X}_{ap}^{H}(\boldsymbol{n})\boldsymbol{w}(\boldsymbol{n})\right] \\ &+ \frac{1}{2}\boldsymbol{\lambda}_{ap}^{T}(n)\left[\boldsymbol{d}_{ap}(n) - \boldsymbol{X}_{ap}^{T}(n)\boldsymbol{w}^{*}(n)\right] \end{aligned} \tag{7.4}$$

However, since (7.2)–(7.4) contain transformations of the input data to a complex RKHS space, Wirtinger's Calculus for RKHS described in references [1, 2] can be applied to evaluate the complex cost function gradient. The rules for the new calculus are generalized for use with general Hilbert spaces. Based on Wirtinger's calculus rules (see Chap. 4), the gradient w.r.t. to $\mathbf{w}^{*}(n)$ was found as follows.

Initially rewriting (7.4) by expanding terms, we get

$$\begin{aligned} F[\boldsymbol{w}(\boldsymbol{n})] =& \frac{1}{2}\boldsymbol{w}(\boldsymbol{n})^{\boldsymbol{H}}[\boldsymbol{w}(\boldsymbol{n}) - \boldsymbol{w}(\boldsymbol{n}-1)] - \frac{1}{2}\boldsymbol{w}(\boldsymbol{n}-1)^{\boldsymbol{H}}[\boldsymbol{w}(\boldsymbol{n}) - \boldsymbol{w}(\boldsymbol{n}-1)] \\ &+ \frac{1}{2}\boldsymbol{\lambda}_{ap}^{H}(\boldsymbol{n})\boldsymbol{d}_{ap}^{*}(\boldsymbol{n}) - \frac{1}{2}\boldsymbol{\lambda}_{ap}^{H}(\boldsymbol{n})\boldsymbol{X}_{ap}^{H}(\boldsymbol{n})\boldsymbol{w}(\boldsymbol{n}) \\ &+ \frac{1}{2}\boldsymbol{\lambda}_{ap}^{T}(n)\boldsymbol{d}_{ap}(n) - \frac{1}{2}\boldsymbol{\lambda}_{ap}^{T}(n)\boldsymbol{X}_{ap}^{T}(n)\boldsymbol{w}^{*}(n) \end{aligned} \tag{7.5}$$

Additional factoring yields

$$\begin{aligned} F[\boldsymbol{w}(\boldsymbol{n})] =& \frac{1}{2}[\boldsymbol{w}(\boldsymbol{n}) - \boldsymbol{w}(\boldsymbol{n}-1)]^{\boldsymbol{T}}\boldsymbol{w}^{*}(\boldsymbol{n}) \\ &- \frac{1}{2}\boldsymbol{w}(\boldsymbol{n}-1)^{\boldsymbol{H}}\boldsymbol{w}(\boldsymbol{n}) + \frac{1}{2}\boldsymbol{w}(\boldsymbol{n}-1)^{\boldsymbol{H}}\boldsymbol{w}(\boldsymbol{n}-1) \\ &+ \frac{1}{2}\boldsymbol{\lambda}_{ap}^{H}(\boldsymbol{n})\boldsymbol{d}_{ap}^{*}(\boldsymbol{n}) - \frac{1}{2}\boldsymbol{\lambda}_{ap}^{H}(\boldsymbol{n})\boldsymbol{X}_{ap}^{H}(\boldsymbol{n})\boldsymbol{w}(\boldsymbol{n}) \\ &+ \frac{1}{2}\boldsymbol{\lambda}_{ap}^{T}(n)\boldsymbol{d}_{ap}(n) - \frac{1}{2}\boldsymbol{\lambda}_{ap}^{T}(n)\boldsymbol{X}_{ap}^{T}(n)\boldsymbol{w}^{*}(n) \end{aligned} \tag{7.6}$$

After reordering terms, we get

$$\begin{aligned} F[\boldsymbol{w}(\boldsymbol{n})] =& \frac{1}{2}[\boldsymbol{w}(\boldsymbol{n}) - \boldsymbol{w}(\boldsymbol{n}-1)]^{\boldsymbol{T}}\boldsymbol{w}^{*}(\boldsymbol{n}) \\ &- \frac{1}{2}\boldsymbol{\lambda}_{ap}^{T}(\boldsymbol{n})\boldsymbol{X}_{ap}^{T}(\boldsymbol{n})\boldsymbol{w}^{*}(\boldsymbol{n}) - \frac{1}{2}\boldsymbol{\lambda}_{ap}^{H}(\boldsymbol{n})\boldsymbol{X}_{ap}^{H}(\boldsymbol{n})\boldsymbol{w}(\boldsymbol{n}) \end{aligned}$$

$$- \frac{1}{2}\boldsymbol{w}(\boldsymbol{n}-1)^{\boldsymbol{H}}\boldsymbol{w}(\boldsymbol{n}) + \frac{1}{2}\boldsymbol{w}(\boldsymbol{n}-1)^{\boldsymbol{H}}\boldsymbol{w}(\boldsymbol{n}-1)$$
$$+ \frac{1}{2}\boldsymbol{\lambda}_{ap}^{H}(n)\boldsymbol{d}_{ap}^{*}(n) + \frac{1}{2}\boldsymbol{\lambda}_{ap}^{T}(n)\boldsymbol{d}_{ap}(n) \tag{7.7}$$

This can be rewritten as

$$\begin{aligned}\boldsymbol{F}[\boldsymbol{w}(\boldsymbol{n})] =& \left(\frac{1}{2}[\boldsymbol{w}(\boldsymbol{n}) - \boldsymbol{w}(\boldsymbol{n}-1)]^{\boldsymbol{T}} - \frac{1}{2}\boldsymbol{\lambda}_{\boldsymbol{ap}}^{\boldsymbol{T}}(\boldsymbol{n})\boldsymbol{X}_{\boldsymbol{ap}}^{\boldsymbol{T}}(\boldsymbol{n})\right)\boldsymbol{w}^{*}(\boldsymbol{n}) \\ &+ \left(-\frac{1}{2}\boldsymbol{w}(\boldsymbol{n}-1)^{\boldsymbol{H}} - \frac{1}{2}\boldsymbol{\lambda}_{\boldsymbol{ap}}^{\boldsymbol{H}}(\boldsymbol{n})\boldsymbol{X}_{\boldsymbol{ap}}^{\boldsymbol{H}}(\boldsymbol{n})\right)\boldsymbol{w}(\boldsymbol{n}) \\ &+ \frac{1}{2}\boldsymbol{w}(n-1)^{H}\boldsymbol{w}(n-1) + \frac{1}{2}\boldsymbol{\lambda}_{ap}^{H}(n)\boldsymbol{d}_{ap}^{*}(n) + \frac{1}{2}\boldsymbol{\lambda}_{ap}^{T}(n)\boldsymbol{d}_{ap}(n)\end{aligned} \tag{7.8}$$

For the gradient with respect to $\mathbf{w}^{*}(n)$, we write

$$\begin{aligned}\nabla_{\boldsymbol{w}^{*}(n)}F[\boldsymbol{w}(n)] =& \nabla_{\boldsymbol{w}^{*}(n)}\left[\left(\frac{1}{2}[\boldsymbol{w}(n) - \boldsymbol{w}(n-1)]^{T} - \frac{1}{2}\boldsymbol{\lambda}_{ap}^{T}(n)\boldsymbol{X}_{ap}^{T}(n)\right)\boldsymbol{w}^{*}(n)\right] \\ &- \nabla_{\boldsymbol{w}^{*}(n)}\left[\left(\frac{1}{2}\boldsymbol{w}(n-1)^{H} + \frac{1}{2}\boldsymbol{\lambda}_{ap}^{H}(n)\boldsymbol{X}_{ap}^{H}(n)\right)\boldsymbol{w}(n)\right] \\ &+ \nabla_{\boldsymbol{w}^{*}(n)}\left[\frac{1}{2}\boldsymbol{w}(n-1)^{H}\boldsymbol{w}(n-1) + \frac{1}{2}\boldsymbol{\lambda}_{ap}^{H}(n)\boldsymbol{d}_{ap}^{*}(n)\right. \\ &\left.+\frac{1}{2}\boldsymbol{\lambda}_{ap}^{T}(n)\boldsymbol{d}_{ap}(n)\right]\end{aligned} \tag{7.9}$$

Considering that, for finding $\nabla_{\boldsymbol{w}^{*}(n)}$, we may consider $\mathbf{w}(n-1)$ and $\mathbf{w}^{*}(n-1)$ fixed with respect to $\mathbf{w}^{*}(n)$ yields

$$\begin{aligned}\nabla_{\boldsymbol{w}^{*}(n)}F[\boldsymbol{w}(n)] =& \nabla_{\boldsymbol{w}^{*}(n)}\left[\left(\frac{1}{2}[\boldsymbol{w}(n) - \boldsymbol{w}(n-1)]^{T} - \frac{1}{2}\boldsymbol{\lambda}_{ap}^{T}(n)\boldsymbol{X}_{ap}^{T}(n)\right)\boldsymbol{w}^{*}(n)\right] \\ &- \nabla_{\boldsymbol{w}^{*}(n)}\left[\left(\frac{1}{2}\boldsymbol{w}(n-1)^{H} + \frac{1}{2}\boldsymbol{\lambda}_{ap}^{H}(n)\boldsymbol{X}_{ap}^{H}(n)\right)\boldsymbol{w}(n)\right]\end{aligned} \tag{7.10}$$

Which we rewrite as (after expanding the first term on the right side of (7.10)):

$$\begin{aligned}\nabla_{\boldsymbol{w}^{*}(n)}F[\boldsymbol{w}(n)] =& \nabla_{\boldsymbol{w}^{*}(n)}\left[\frac{1}{2}\boldsymbol{w}(n)^{T}\boldsymbol{w}^{*}(n)\right] \\ &- \nabla_{\boldsymbol{w}^{*}(n)}\left[\left(\frac{1}{2}\boldsymbol{w}(n-1)^{T} + \frac{1}{2}\boldsymbol{\lambda}_{ap}^{T}(n)\boldsymbol{X}_{ap}^{T}(n)\right)\boldsymbol{w}^{*}(n)\right] \\ &- \nabla_{\boldsymbol{w}^{*}(n)}\left[\left(\frac{1}{2}\boldsymbol{w}(n-1)^{H} + \frac{1}{2}\boldsymbol{\lambda}_{ap}^{H}(n)\boldsymbol{X}_{ap}^{H}(n)\right)\boldsymbol{w}(n)\right]\end{aligned} \tag{7.11}$$

Since $\mathbf{w}(n)$, $\mathbf{X}_{ap}(n)$ are mapped to the RKHS space, we can rewrite (7.11) using inner products, which yields

$$\nabla_{\boldsymbol{w}^*(\boldsymbol{n})}\boldsymbol{F}[\boldsymbol{w}(\boldsymbol{n})] = \frac{1}{2}\nabla_{\boldsymbol{w}^*(\boldsymbol{n})}\big[< \boldsymbol{w}(\boldsymbol{n}), \boldsymbol{w}(\boldsymbol{n}) >_{\mathbb{H}} - < \boldsymbol{a}, \boldsymbol{w}(\boldsymbol{n}) >_{\mathbb{H}} - < \boldsymbol{b}, \boldsymbol{w}^*(\boldsymbol{n}) >_{\mathbb{H}}\big]$$

where $\boldsymbol{a} = \frac{1}{2}\boldsymbol{w}(n-1) + \frac{1}{2}\boldsymbol{X}_{ap}(n)\boldsymbol{\lambda}_{ap}(n)$, $\boldsymbol{b} = \frac{1}{2}\boldsymbol{w}^*(n-1) + \frac{1}{2}\boldsymbol{X}^*_{ap}(n)\boldsymbol{\lambda}^*_{ap}(n)$.

Finally, using the gradient properties in Chap. 4, we get

$$\nabla_{\boldsymbol{w}^*(n)}F[\boldsymbol{w}(n)] = \frac{1}{2}[\boldsymbol{w}(n) - \boldsymbol{w}(n-1)] - \frac{1}{2}\boldsymbol{X}_{ap}(n)\boldsymbol{\lambda}_{ap}(n) \tag{7.12}$$

which is the gradient result.

Obtaining Eq. (7.12) also involves differentiation with respect to a vector, since **a**, **b** and $\mathbf{w}(n)$ may be vectors. A description of the process can be found in [5], and the interested reader can verify that the steps used in obtaining (7.12) follow the rules described there.

A key point to note, however, is that the application of the rules of the modified Wirtinger's calculus for Hilbert spaces from [1] allows for evaluation of the gradient of the cost function defined on a complex RKHS. Concerns regarding analyticity (or Fréchet complex analyticity for functions on a complex RKHS—see Chap. 4) are not a hindrance for evaluating the gradient when considering the use of Wirtinger's calculus.

Setting the gradient equal to zero yields:

$$0 = \frac{1}{2}[\boldsymbol{w}(\boldsymbol{n}) - \boldsymbol{w}(\boldsymbol{n}-1)] \quad - \quad \frac{1}{2}\boldsymbol{X}_{\boldsymbol{ap}}(\boldsymbol{n})\boldsymbol{\lambda}_{\boldsymbol{ap}}(\boldsymbol{n})$$

$$\boldsymbol{w}(n) - \boldsymbol{w}(n-1) \quad = \quad \boldsymbol{X}_{ap}(n)\boldsymbol{\lambda}_{ap}(n) \tag{7.13}$$

Therefore,

$$\boldsymbol{w}(n) = \boldsymbol{w}(n-1) + \boldsymbol{X}_{ap}(n)\boldsymbol{\lambda}_{ap}(n) \tag{7.14}$$

Substituting into the constraint $\boldsymbol{d}_{ap}(n) - \boldsymbol{X}^T_{ap}(n)\boldsymbol{w}^*(n) = 0$ yields

$$\begin{aligned}
&\boldsymbol{d}_{ap}(n) - \boldsymbol{X}^T_{ap}(n)[\boldsymbol{w}(n-1) + \boldsymbol{X}_{ap}(n)\boldsymbol{\lambda}_{ap}(n)]^* = 0\\
&\boldsymbol{d}_{ap}(n) - \boldsymbol{X}^T_{ap}(n)\boldsymbol{w}^*(n-1) - \boldsymbol{X}^T_{ap}(n)(\boldsymbol{X}_{ap}(n)\boldsymbol{\lambda}_{ap}(n))^* = 0\\
&\boldsymbol{d}_{ap}(n) - \boldsymbol{X}^T_{ap}(n)\boldsymbol{w}^*(n-1) = (\boldsymbol{X}^H_{ap}(n)\boldsymbol{X}_{ap}(n)\boldsymbol{\lambda}_{ap}(n))^*
\end{aligned} \tag{7.15}$$

And since the estimation error, $e_{ap}(n)$, is defined as

$$e_{ap}(n) = \boldsymbol{d}_{ap}(n) - \boldsymbol{X}^T_{ap}(n)\boldsymbol{w}^*(n-1) \tag{7.16}$$

Therefore, combining (7.15), (7.16) yields

$$e_{ap}(n) = (\boldsymbol{X}_{ap}^H(n)\boldsymbol{X}_{ap}(n)\boldsymbol{\lambda}_{ap}(n))^*$$
$$\boldsymbol{\lambda}_{ap}(n) = \left(\boldsymbol{X}_{ap}^H(n)\boldsymbol{X}_{ap}(n)\right)^{-1} e_{ap}^*(n) \tag{7.17}$$

And the algorithm weight update can be written as

$$\boldsymbol{w}(n) = \boldsymbol{w}(n-1) + \mu\boldsymbol{X}_{ap}(n)\left(\boldsymbol{X}_{ap}^H(n)\boldsymbol{X}_{ap}(n)\right)^{-1} e_{ap}^*(n) \tag{7.18}$$

where $\mu$ is chosen considering stability, convergence rate and misadjustment.

We also derive alternative forms of CKAPA based on the use of the stochastic gradient and Newton's method, as well as their regularized versions. Details are in Sect. 7.2, resulting in four CKAPA-type algorithms. Equation (7.18) is the same as one of these algorithms (specifically CKAPA-2, shown later).

For finite complexity, the algorithm should be cast in terms of inner products in the RKHS, where the kernel trick can be used. As in the approach of [4], we assume the initial weights are $\mathbf{w}(0) = 0$, and use (7.18) recursively to find $\mathbf{w}(n-1)$, which becomes

$$\begin{gathered}\boldsymbol{w}(0) = 0\\ \boldsymbol{w}(1) = \boldsymbol{\mu\Phi}(z(1))\cdot < \boldsymbol{\Phi}(z(1)), \boldsymbol{\Phi}(z(1)) >_S^{-1} \cdot\boldsymbol{d}^*(1) = \boldsymbol{a}_1(1)\boldsymbol{\Phi}(z(1))\\ \vdots\\ \boldsymbol{w}(n-1) = \sum_{j=1}^{n-1} a_j(n-1)\boldsymbol{\Phi}(z(j))\end{gathered} \tag{7.19}$$

where $\boldsymbol{\Phi}(z(n))$ is the input in the RKHS, $< \boldsymbol{\Phi}(z(p)), \boldsymbol{\Phi}(z(q)) >_S$ represents the inner product $\boldsymbol{\Phi}(z(q))^H \boldsymbol{\Phi}(z(p))$ in the RKHS (either complexified or complex), and $a_j(n-1)$ is the expansion coefficient to apply to the transform mapping $\boldsymbol{\Phi}(z(j))$.

The algorithm output (i.e. estimates) can then be found using

$$\hat{\boldsymbol{d}}_{ap}(\boldsymbol{n}) = \boldsymbol{X}_{ap}^T(\boldsymbol{n})\boldsymbol{w}^*(\boldsymbol{n}-1) = \begin{bmatrix}\sum_{j=1}^{n-1} a_j^*(n-1) < \boldsymbol{\Phi}(z(n-K+1)), \boldsymbol{\Phi}(z(j)) >_S\\ \vdots\\ \sum_{j=1}^{n-1} a_j^*(n-1) < \boldsymbol{\Phi}(z(n-1)), \boldsymbol{\Phi}(z(j)) >_S\\ \sum_{j=1}^{n-1} a_j^*(n-1) < \boldsymbol{\Phi}(z(n)), \boldsymbol{\Phi}(z(j)) >_S\end{bmatrix} \tag{7.20}$$

with estimation error, $\boldsymbol{e}_{ap}(n) = \boldsymbol{d}_{ap}(n) - \hat{\boldsymbol{d}}_{ap}(n)$.

The expansion coefficients can be found as follows. Rewriting the recursion of (7.19) with added details.

For $\mathbf{w}(1)$ (assuming $\mathbf{w}(0) = 0$):

$$\begin{aligned} \boldsymbol{w}(1) &= \boldsymbol{w}(0) + \mu \boldsymbol{X}_{ap}(1)\left(\boldsymbol{X}_{ap}^{H}(1)\boldsymbol{X}_{ap}(1)\right)^{-1}\boldsymbol{e}_{ap}^{*}(1) \\ &= \mu \boldsymbol{\Phi}(z(1)) < \boldsymbol{\Phi}(z(1)), \boldsymbol{\Phi}(z(1)) >_{S}^{-1} d^{*}(1) \\ &= \left(\mu < \boldsymbol{\Phi}(z(1)), \boldsymbol{\Phi}(z(1)) >_{S}^{-1} d^{*}(1)\right)\boldsymbol{\Phi}(z(1)) \\ &= a_1(1)\boldsymbol{\Phi}(z(1)) \end{aligned} \tag{7.21}$$

where $a_k(i)$ refers to the scaling applied to vector $\boldsymbol{\Phi}(z(k))$ at iteration $i$ to find $\mathbf{w}(i)$.

For $\mathbf{w}(2)$:

$$\begin{aligned} \boldsymbol{w}(2) &= \boldsymbol{w}(1) + \mu \boldsymbol{X}_{ap}(2)\left(\boldsymbol{X}_{ap}^{H}(2)\boldsymbol{X}_{ap}(2)\right)^{-1}\boldsymbol{e}_{ap}^{*}(2) \\ &= a_1(1)\boldsymbol{\Phi}(z(1)) + \mu \boldsymbol{X}_{ap}(2)\tilde{\boldsymbol{e}}(2) \\ &= a_1(1)\boldsymbol{\Phi}(z(1)) + [\boldsymbol{\Phi}(z(1))\boldsymbol{\Phi}(z(2))][\tilde{\boldsymbol{e}}_1(2)\tilde{\boldsymbol{e}}_2(2)]^{T} \\ &= a_1(1)\boldsymbol{\Phi}(z(1)) + \tilde{\boldsymbol{e}}_1(2)\boldsymbol{\Phi}(z(1)) + \tilde{\boldsymbol{e}}_2(2)\boldsymbol{\Phi}(z(2)) \\ &= (a_1(1) + \tilde{\boldsymbol{e}}_1(2))\boldsymbol{\Phi}(z(1)) + \tilde{\boldsymbol{e}}_2(2)\boldsymbol{\Phi}(z(2)) \\ &= a_1(2)\boldsymbol{\Phi}(z(1)) + a_2(2)\boldsymbol{\Phi}(z(2)) \end{aligned} \tag{7.22}$$

where

$$\tilde{\boldsymbol{e}}(n) = \left(\boldsymbol{X}_{ap}^{H}(n)\boldsymbol{X}_{ap}(n)\right)^{-1}\boldsymbol{e}_{ap}^{*}(n) \tag{7.23}$$

and $\tilde{\boldsymbol{e}}_j(n)$ denotes the element in location $j$ of $\tilde{\boldsymbol{e}}(n)$.

Equations (7.21), (7.22) show the $a_k(i)$ coefficients for $\mathbf{w}(1)$, $\mathbf{w}(2)$. For a general approach for $\mathbf{w}(n)$, we consider expressing the weight update (7.18) with the following method (from [4]).

Based on (7.21), (7.22) and the Representer Theorem, we can express $\mathbf{w}(n-1)$ as

$$\boldsymbol{w}(n-1) = \sum_{j=1}^{n-1} a_j(n-1)\boldsymbol{\Phi}(z(j)) \tag{7.24}$$

Now we rewrite the weight update (7.18) as

$$\boldsymbol{w}(n) = \boldsymbol{w}(n-1) + \mu \boldsymbol{X}_{ap}(n)\tilde{\boldsymbol{e}}(n) \tag{7.25}$$

where

$$\tilde{\boldsymbol{e}}(n) = \left(\boldsymbol{X}_{ap}^{H}(n)\boldsymbol{X}_{ap}(n) + \varepsilon \boldsymbol{I}\right)^{-1}\boldsymbol{e}_{ap}^{*}(n) \tag{7.26}$$

Combining (7.24) and (7.25), we get

$$\boldsymbol{w}(n) = \sum_{j=1}^{n-1} a_j(n-1)\boldsymbol{\Phi}(z(j)) + \mu \boldsymbol{X}_{ap}(n)\tilde{\boldsymbol{e}}(n). \tag{7.27}$$

which can be expanded as

$$\boldsymbol{w}(n) = \sum_{j=1}^{n-1} a_j(n-1)\boldsymbol{\Phi}(z(j)) + \mu \sum_{k=1}^{K} \tilde{\boldsymbol{e}}_k(n)\boldsymbol{\Phi}(z(n-K+k)) \tag{7.28}$$

where $\tilde{\boldsymbol{e}}_k(n)$ is the k-th element of vector $\tilde{\boldsymbol{e}}(n)$.

From Eq. (7.28), we can see that only the coefficients $a_{n-k+1}$ to $a_{n-1}$ are modified from iteration $n-1$ to $n$. Therefore, the update is

$$a_k(i) = \begin{cases} \mu \tilde{\boldsymbol{e}}_K(i), & k = i \\ a_k(i-1) + \mu \tilde{\boldsymbol{e}}_{K+k-i}(i), & i-K+1 \le k \le i-1 \\ a_k(i-1), & 1 \le k < i-K+1 \end{cases} \tag{7.29}$$

can be used for finding the $a_j(n)$ coefficients.

Summarizing the estimates, we get

$$\hat{\boldsymbol{d}}_{ap}(n) = \boldsymbol{X}_{ap}^T(n)\boldsymbol{w}^*(n-1) = \begin{bmatrix} \sum_{j=1}^{n-1} a_j^*(n-1) < \boldsymbol{\Phi}(z(n-K+1)), \boldsymbol{\Phi}(z(j)) >_S \\ \vdots \\ \sum_{j=1}^{n-1} a_j^*(n-1) < \boldsymbol{\Phi}(z(n-1)), \boldsymbol{\Phi}(z(j)) >_S \\ \sum_{j=1}^{n-1} a_j^*(n-1) < \boldsymbol{\Phi}(z(n)), \boldsymbol{\Phi}(z(j)) >_S \end{bmatrix} \tag{7.30}$$

with estimation error, $\boldsymbol{e}_{ap}(n) = \boldsymbol{d}_{ap}(n) - \hat{\boldsymbol{d}}_{ap}(n)$, and expansion coefficients, $a_k(i)$, shown previously.

Equations (7.29) and (7.30) are one of the CKAPA forms derived, where the inner products may be defined on either a complexified or complex RKHS space. These algorithms are equivalent to the CKAPA-2a and CKAPA-2b algorithms, respectively, of the following section.

## 7.2 Derivation of Alternative Methods

For alternative forms of the CKAPA algorithm, we consider the approach of [4] for deriving KAPA forms, and extend it for complex data based on the modified Wirtinger's calculus.

From Chap. 2, the APA forms shown were

$$\boldsymbol{w}(n) = \boldsymbol{w}(n-1) + \mu \boldsymbol{X}(n)\left[\boldsymbol{d}^*(n) - \boldsymbol{X}^H(n)\boldsymbol{w}(n-1)\right] \tag{7.31}$$

$$\boldsymbol{w}(n) = \boldsymbol{w}(n-1) + \mu \boldsymbol{X}(n)[\boldsymbol{X}^H(n)\boldsymbol{X}(n) + \varepsilon \boldsymbol{I}]^{-1}\left[\boldsymbol{d}^*(n) - \boldsymbol{X}^H(n)\boldsymbol{w}(n-1)\right] \tag{7.32}$$

$$\boldsymbol{w}(n) = (1-\eta\lambda)\boldsymbol{w}(n-1) + \eta \boldsymbol{X}(n)\left[\boldsymbol{d}^*(n) - \boldsymbol{X}^H(n)\boldsymbol{w}(n-1)\right] \tag{7.33}$$

$$\boldsymbol{w}(n) = (1-\eta)\boldsymbol{w}(n-1) + \eta \boldsymbol{X}(n)[\lambda \boldsymbol{I} + \boldsymbol{X}^H(n)\boldsymbol{X}(n)]^{-1}\boldsymbol{d}^*(n) \tag{7.34}$$

which were referred to as APA-1 to APA-4, respectively. Equation (7.32) is the APA form followed in Sect. 7.1 and is based on Newton's recursion. Meanwhile, Eqs. (7.31), (7.33) and (7.34) are alternative forms based on: (1) Steepest descent, (2) Regularized least-squares and (3) Regularized least-squares with Newton's recursion, respectively.

To convert these algorithms to complex kernel methods, the input vectors forming matrix $\mathbf{X}(n)$ are transformed to either a complex or complexified RKHS.

The cost functions needed to derive these algorithms are (from Chap. 2)

$$\min_{\boldsymbol{w}} E\left|d(n) - \boldsymbol{w}^H(n)\boldsymbol{\Phi}(\boldsymbol{x}(n))\right|^2 \tag{7.35}$$

and

$$\min_{\boldsymbol{w}} E\left|d(n) - \boldsymbol{w}^H(n)\boldsymbol{\Phi}(\boldsymbol{x}(n))\right|^2 + \lambda \boldsymbol{w}(n)^2 \tag{7.36}$$

However, since (7.35) and (7.36) contain transformations of the input data to a complex RKHS space, Wirtinger's Calculus for RHKS described in Chap. 4 should be considered to evaluate the cost function gradient.

Initially writing (7.35) as

$$\begin{aligned} \boldsymbol{F}[\boldsymbol{w}(\boldsymbol{n})] =& \boldsymbol{E}\left[\left(\boldsymbol{d}(\boldsymbol{n}) - \boldsymbol{w}^{\boldsymbol{H}}(\boldsymbol{n})\boldsymbol{\Phi}(\boldsymbol{x}(\boldsymbol{n}))\right)^{\boldsymbol{H}}\left(\boldsymbol{d}(\boldsymbol{n}) - \boldsymbol{w}^{\boldsymbol{H}}(\boldsymbol{n})\boldsymbol{\Phi}(\boldsymbol{x}(\boldsymbol{n}))\right)\right] \\ =& \boldsymbol{E}[|\boldsymbol{d}(\boldsymbol{n})|^2 - \boldsymbol{d}^*(\boldsymbol{n})\boldsymbol{w}^{\boldsymbol{H}}(\boldsymbol{n})\boldsymbol{\Phi}(\boldsymbol{x}(\boldsymbol{n})) - \boldsymbol{d}(\boldsymbol{n})\boldsymbol{\Phi}^{\boldsymbol{H}}(\boldsymbol{x}(\boldsymbol{n}))\boldsymbol{w}(\boldsymbol{n}) \\ &+ \boldsymbol{\Phi}^H(\boldsymbol{x}(n))\boldsymbol{w}(n)\boldsymbol{w}^H(n)\boldsymbol{\Phi}(\boldsymbol{x}(n))] \end{aligned} \tag{7.37}$$

For the gradient with respect to $\mathbf{w}^*(n)$, we can write

$$\begin{aligned}\nabla_{w^*(n)}F[w(n)] =& E\big[\nabla_{w^*(n)}\big[|d(n)|^2\big] \\ & - \nabla_{w^*(n)}\big[d^*(n)\boldsymbol{\Phi}^T(\boldsymbol{x}(n))\boldsymbol{w}^*(n)\big] \\ & - \nabla_{w^*(n)}\big[d(n)\boldsymbol{\Phi}^H(\boldsymbol{x}(n))\boldsymbol{w}(n)\big] \\ & + \nabla_{w^*(n)}\big[\boldsymbol{\Phi}^H(\boldsymbol{x}(n))\boldsymbol{w}(n)\boldsymbol{\Phi}^T(\boldsymbol{x}(n))\boldsymbol{w}^*(n)\big]\big] \end{aligned} \tag{7.38}$$

since $\mathbf{a}^H\mathbf{b} = \mathbf{b}^T\mathbf{a}^*$ when **a**, **b** are column vectors of the same length (i.e. the result is scalar).

Considering that, for finding $\nabla_{w^*(n)}$, we may consider $d(n)$ fixed with respect to $\mathbf{w}^*(n)$ yields

$$\begin{aligned}&\nabla_{w^*(n)}F[w(n)] = E[\nabla_{w^*(n)}\big[\boldsymbol{\Phi}^H(\boldsymbol{x}(n))\boldsymbol{w}(n)\boldsymbol{\Phi}^T(\boldsymbol{x}(n))\boldsymbol{w}^*(n)\big] \\ & - d^*(n)\cdot\nabla_{w^*(n)}\big[\boldsymbol{\Phi}^T(\boldsymbol{x}(n))\boldsymbol{w}^*(n)\big] - d(n)\cdot\nabla_{w^*(n)}\big[\boldsymbol{\Phi}^H(\boldsymbol{x}(n))\boldsymbol{w}(n)\big]\big]\end{aligned} \tag{7.39}$$

Since $\mathbf{w}(n)$, $\boldsymbol{\Phi}^T(\boldsymbol{x}(n))$ are vectors in the RKHS space, we can rewrite (7.39) using inner products, which yields

$$\begin{aligned}\nabla_{w^*(n)}F[w(n)] =& E\big[\nabla_{w^*(n)}\big[< \boldsymbol{\Phi}^*(\boldsymbol{x}(n)), \boldsymbol{w}^*(n) >_{\mathbb{H}} < \boldsymbol{\Phi}(\boldsymbol{x}(n)), \boldsymbol{w}(n) >_{\mathbb{H}}\big] \\ & - \boldsymbol{d}^*(\boldsymbol{n})\cdot\nabla_{w^*(n)}[< \boldsymbol{\Phi}(\boldsymbol{x}(n)), \boldsymbol{w}(n) >_{\mathbb{H}}] \\ & - d(n)\cdot\nabla_{w^*(n)}\big[< \boldsymbol{\Phi}^*(\boldsymbol{x}(n)), \boldsymbol{w}^*(n) >_{\mathbb{H}}\big]\big]\end{aligned} \tag{7.40}$$

After using the gradient properties from Chap. 4, (7.40) becomes

$$\begin{aligned}\nabla_{w^*(n)}F[w(n)] =& E\big[< \boldsymbol{\Phi}^*(\boldsymbol{x}(n)), \boldsymbol{w}^*(n) >_{\mathbb{H}} \nabla_{w^*(n)}[< \boldsymbol{\Phi}(\boldsymbol{x}(n)), \boldsymbol{w}(n) >_{\mathbb{H}}] \\ & - \boldsymbol{d}^*(\boldsymbol{n})\cdot\nabla_{w^*(n)}[< \boldsymbol{\Phi}(\boldsymbol{x}(n)), \boldsymbol{w}(n) >_{\mathbb{H}}] \\ & - \boldsymbol{d}(\boldsymbol{n})\cdot\nabla_{w^*(n)}\big[< \boldsymbol{\Phi}^*(\boldsymbol{x}(n)), \boldsymbol{w}^*(n) >_{\mathbb{H}}\big]\big] \\ =& E\big[< \boldsymbol{\Phi}^*(\boldsymbol{x}(n)), \boldsymbol{w}^*(n) >_{\mathbb{H}} \boldsymbol{\Phi}(\boldsymbol{x}(n))\big] \\ & - E\big[d^*(n)\cdot\boldsymbol{\Phi}(\boldsymbol{x}(n))\big]\end{aligned} \tag{7.41}$$

(since $\boldsymbol{\Phi}(\boldsymbol{x}(n))$,$\boldsymbol{\Phi}^*(\boldsymbol{x}(n))$ are considered fixed w.r.t. $\mathbf{w}^*(n)$).

Note that $\boldsymbol{\Phi}^T(\boldsymbol{x}(n))$ and $\mathbf{w}(n)$ are vectors. Therefore, obtaining Eq. (7.41) also involves differentiation with respect to a vector. Additionally, as noted before, the application of Wirtinger's calculus for Hilbert spaces allows for the gradient of the cost function defined on a complex RKHS to be found here.

To find the optimal weights, $\mathbf{w}^o$, we set the gradient (7.41) to zero to find the minimum of the cost function (7.35).

Rewriting Eq. (7.41) as

$$\begin{aligned}\nabla_{w^*(n)}F[w(n)] &= E\big[\boldsymbol{\Phi}(\boldsymbol{x}(n)) < \boldsymbol{\Phi}^*(\boldsymbol{x}(n)), \boldsymbol{w}^*(n) >_{\mathbb{H}}\big] - E\big[d^*(n)\cdot\boldsymbol{\Phi}(\boldsymbol{x}(n))\big] \\ &= E\big[\boldsymbol{\Phi}(\boldsymbol{x}(n))\boldsymbol{w}^T(n)\boldsymbol{\Phi}^*(\boldsymbol{x}(n))\big] - E\big[d^*(n)\cdot\boldsymbol{\Phi}(\boldsymbol{x}(n))\big]\end{aligned} \tag{7.42}$$

And setting to zero yields

$$
\begin{aligned}
0 &= E\left[\boldsymbol{\Phi}(\boldsymbol{x}(n))\boldsymbol{w}^T(n)\boldsymbol{\Phi}^*(\boldsymbol{x}(n))\right] - E\left[d^*(n)\cdot\boldsymbol{\Phi}(\boldsymbol{x}(n))\right]\\
0 &= E\left[\boldsymbol{\Phi}(\boldsymbol{x}(n))\boldsymbol{\Phi}^H(\boldsymbol{x}(n))\boldsymbol{w}(n)\right] - E\left[d^*(n)\cdot\boldsymbol{\Phi}(\boldsymbol{x}(n))\right]\\
&E\left[\boldsymbol{\Phi}(\boldsymbol{x}(n))\boldsymbol{\Phi}^H(\boldsymbol{x}(n))\boldsymbol{w}(n)\right] = E\left[d^*(n)\cdot\boldsymbol{\Phi}(\boldsymbol{x}(n))\right]\\
&E\left[\boldsymbol{\Phi}(\boldsymbol{x}(n))\boldsymbol{\Phi}^H(\boldsymbol{x}(n))\right]E[\boldsymbol{w}(n)] = E\left[d^*(n)\cdot\boldsymbol{\Phi}(\boldsymbol{x}(n))\right]
\end{aligned}
\tag{7.43}
$$

Therefore, we obtain

$$
\boldsymbol{w}^o = E[\boldsymbol{w}(n)] = E\left[\boldsymbol{\Phi}(\boldsymbol{x}(n))\boldsymbol{\Phi}^H(\boldsymbol{x}(n))\right]^{-1}E\left[d^*(n)\cdot\boldsymbol{\Phi}(\boldsymbol{x}(n))\right] \tag{7.44}
$$

which can clearly be seen to match the Wiener solution form shown after Eq. (2.19) (i.e. $\mathbf{w}^o = \mathbf{R_x}^{-1}\mathbf{r}_{\mathrm{dx}}$, where now $\mathbf{R_x} = E\left[\boldsymbol{\Phi}(\boldsymbol{x}(n))\boldsymbol{\Phi}^H(\boldsymbol{x}(n))\right]$, $\mathbf{r}_{\mathrm{dx}} = E[d^*(n)\cdot\boldsymbol{\Phi}(\boldsymbol{x}(n))]$). The same approach of using the $K$ most recent inputs for improving the estimates of the covariance matrix $\mathbf{R_x}$ and cross-covariance vector $\mathbf{r}_{\mathrm{dx}}$ can be used.

Based on this, and forming a matrix of the transformed input vectors:

$$
\boldsymbol{\Phi}_{ap}(n) = [\boldsymbol{\Phi}(\boldsymbol{x}(n-K+1))\ \boldsymbol{\Phi}(\boldsymbol{x}(n-K+2))\ldots\ \boldsymbol{\Phi}(\boldsymbol{x}(n))] \tag{7.45}
$$

The complex kernel-based versions of APA-1 to APA-4 become

$$
\boldsymbol{w}(n) = \boldsymbol{w}(n-1) + \mu\boldsymbol{\Phi}_{ap}(n)\left[\boldsymbol{d}^*(n) - \boldsymbol{\Phi}_{ap}^H(n)\boldsymbol{w}(n-1)\right] \tag{7.46}
$$

$$
\boldsymbol{w}(\boldsymbol{n}) = \boldsymbol{w}(\boldsymbol{n}-1) + \boldsymbol{\mu}\boldsymbol{\Phi}_{ap}(\boldsymbol{n})[\boldsymbol{\Phi}_{ap}^H(\boldsymbol{n})\boldsymbol{\Phi}_{ap}(\boldsymbol{n}) + \boldsymbol{\varepsilon I}]^{-1}\cdot
$$

$$
\left[\boldsymbol{d}^*(n) - \boldsymbol{\Phi}_{ap}^H(n)\boldsymbol{w}(n-1)\right] \tag{7.47}
$$

$$
\boldsymbol{w}(n) = (1-\eta\lambda)\boldsymbol{w}(n-1) + \eta\boldsymbol{\Phi}_{ap}(n)\left[\boldsymbol{d}^*(n) - \boldsymbol{\Phi}_{ap}^H(n)\boldsymbol{w}(n-1)\right] \tag{7.48}
$$

$$
\boldsymbol{w}(n) = (1-\eta)\boldsymbol{w}(n-1) + \eta\boldsymbol{\Phi}_{ap}(n)[\lambda\boldsymbol{I} + \boldsymbol{\Phi}_{ap}^H(n)\boldsymbol{\Phi}_{ap}(n)]^{-1}\boldsymbol{d}^*(n) \tag{7.49}
$$

Equations (7.46) to (7.49) will be referred to as CKAPA-1 to CKAPA-4.

For finite complexity, the algorithm should be cast in terms of inner products in the RKHS, where the kernel trick can be used. Following the approach in [4], we assume the initial weights are $\mathbf{w}(0) = 0$, and recursively solve for $\mathbf{w}(n-1)$. This may be done for each of (7.46) to (7.49) to find implementations for CKAPA-1 to CKAPA-4.

The CKAPA-2 estimates were shown at the end of Sect. 7.1. Summarizing these, the estimates are

$$
\hat{\boldsymbol{d}}_{ap}(\boldsymbol{n}) = \boldsymbol{\Phi}_{ap}^T(\boldsymbol{n})\boldsymbol{w}^*(\boldsymbol{n}-1) =
$$

$$\begin{bmatrix} \sum_{j=1}^{n-1} a_j^*(n-1) < \boldsymbol{\Phi}(z(n-K+1)), \boldsymbol{\Phi}(z(j)) >_S \\ \vdots \\ \sum_{j=1}^{n-1} a_j^*(n-1) < \boldsymbol{\Phi}(z(n-1)), \boldsymbol{\Phi}(z(j)) >_S \\ \sum_{j=1}^{n-1} a_j^*(n-1) < \boldsymbol{\Phi}(z(n)), \boldsymbol{\Phi}(z(j)) >_S \end{bmatrix} \tag{7.50}$$

with estimation error

$$\boldsymbol{e}_{ap}(n) = \boldsymbol{d}_{ap}(n) - \hat{\boldsymbol{d}}_{ap}(n) \tag{7.51}$$

and expansion coefficients, $a_k(i)$:

$$a_k(i) = \begin{cases} \mu\tilde{\boldsymbol{e}}_K(i), k = i \\ a_k(i-1) + \mu\tilde{\boldsymbol{e}}_{K+k-i}(i), i-K+1 \le k \le i-1 \\ a_k(i-1), 1 \le k < i-K+1 \end{cases} \tag{7.52}$$

where $\tilde{\boldsymbol{e}}_k(i)$ is the k-th element of vector: $\tilde{\boldsymbol{e}}(i) = \left(\boldsymbol{X}_{ap}^H(i)\boldsymbol{X}_{ap}(i) + \varepsilon\boldsymbol{I}\right)^{-1} \boldsymbol{e}_{ap}^*(i)$.

Note that CKAPA-1 is the same, except that the normalization does not occur (i.e. $\tilde{\boldsymbol{e}}(i) = \boldsymbol{e}_{ap}^*(i)$ is used instead). A similar process can be used to find the CKAPA-3, CKAPA-4 algorithms.

## 7.3 Simulation Results

The simulations used to evaluate the performance of the CKAPA-2 algorithms (i.e. CKAPA-2a for the complexified, CKAPA-2b for the complex kernel) are described here. For all tests, the following nonlinear channel equalization task was evaluated.

The nonlinear channel consists of the linear filter:

$$\boldsymbol{t}(\boldsymbol{n}) = (-0.9 + 0.8\boldsymbol{i}) \cdot \boldsymbol{s}(\boldsymbol{n}) + (0.6 - 0.7\boldsymbol{i}) \cdot \boldsymbol{s}(\boldsymbol{n} - 1)$$

followed by the memoryless nonlinearity:

$$\boldsymbol{q}(\boldsymbol{n}) = \boldsymbol{t}(\boldsymbol{n}) + (0.1 - 0.15\boldsymbol{i}) \cdot \boldsymbol{t}^2(\boldsymbol{n}) + (0.06 + 0.05\boldsymbol{i}) \cdot \boldsymbol{t}^3(\boldsymbol{n})$$

The signal, $r(n)$, at the receiver is corrupted by white Gaussian noise, with a signal-to-noise ratio of 16 dB. The input signal generated used the form:

$$\boldsymbol{s}(\boldsymbol{n}) = 0.7\left(\sqrt{1-\boldsymbol{p}^2}\boldsymbol{X}(\boldsymbol{n}) + \boldsymbol{i} \cdot \boldsymbol{\rho}\boldsymbol{Y}(\boldsymbol{n})\right)$$

where $X(n)$, $Y(n)$ are Gaussian random variables. The value of $\rho$ used was 0.1, where a value of $\sqrt{2}/2$ generates circular input, and values close to either 0 or 1 generate highly non-circular input.

The performance of CKAPA-2a, CKAPA-2b in obtaining the channel inverse filter is shown. The tests performed evaluate the following:

Fig. 7.1: CKAPA-2a, using real Gaussian kernel

Fig. 7.2: CKAPA-2a, using real Polynomial kernel

Fig. 7.3: CKAPA-2b, using complex Gaussian kernel

Fig. 7.4: CKAPA-2b, using Bergman kernel (complex)

CKAPA-2a forms a complexified RKHS space from the real kernel specified (see Chap. 4 or [1]). For CKAPA-2b, pure complex kernels are used.

The following kernels were tested [1, 4]:

Real Gaussian Kernel: $\kappa(\boldsymbol{x}, \boldsymbol{y}) = \exp(-\sigma \sum_{i=1}^{N} (x_i - y_i)^2)$.

Real Polynomial Kernel: $\kappa(\boldsymbol{x}, \boldsymbol{y}) = (1 + \boldsymbol{x}^T \boldsymbol{y})^q$.

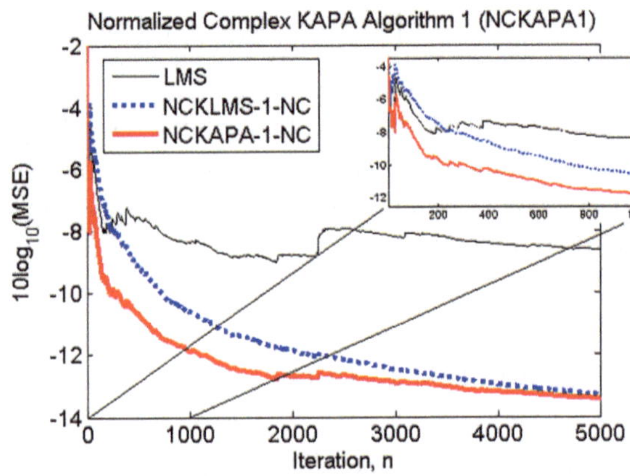


**Fig. 7.1** CKAPA-2a, using Gaussian kernel ($\sigma = 0.01$)

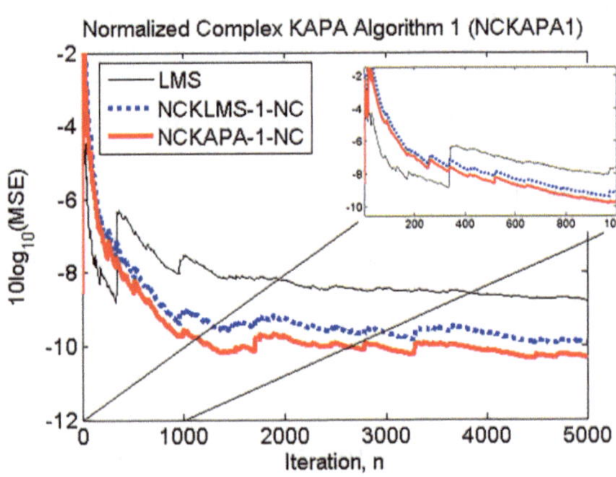


**Fig. 7.2** CKAPA-2a, using Polynomial kernel ($q = 2$)

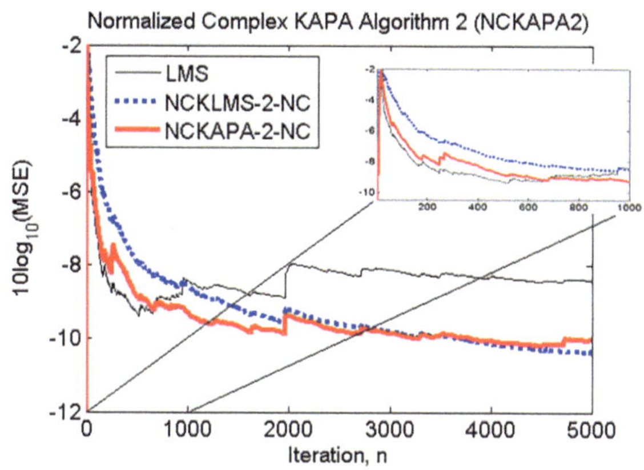


**Fig. 7.3** CKAPA-2b, complex Gaussian kernel ($\sigma = 0.01$)

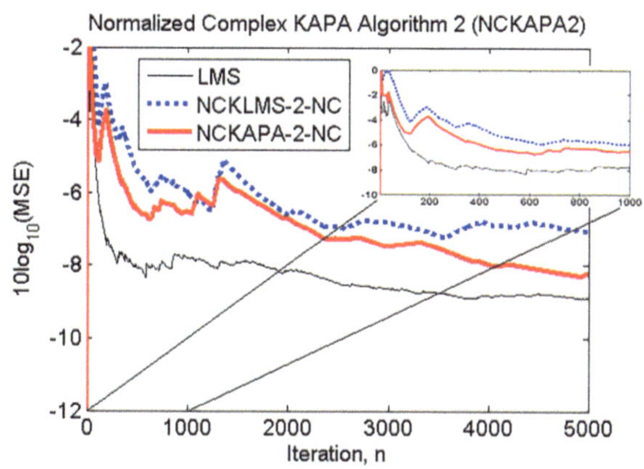


**Fig. 7.4** CKAPA-2b, using Bergman kernel ($q = -2$)

Complex Gaussian Kernel: $\kappa(z, w) = \exp(-\sigma \sum_{i=1}^{N}(z_i - w_i^*)^2)$.

Bergman Kernel: $\kappa(z, w) = \prod_{i=1}^{N}(1 - w_i^* z_i)^q$.

Note the conjugation of $w_i$ in the complex forms. For all kernels, $x, y \in \mathbb{R}^N$, $z, w \in \mathbb{C}^N$ and $\sigma$, $q$ are kernel parameters affecting the transform mapping. The values of $\sigma$, $q$ used are mentioned in the label of each of Figs. 7.1, 7.2, 7.3, and 7.4.

From Figs. 7.1 and 7.2, we see that the performance of the CKAPA-2 provides a considerable convergence rate benefit compared to CKLMS. This is a known advantage for APA versus LMS, and may be of particular value for nonlinear channel modeling and equalization tasks, where the weights in the transform space are often infinite dimensional.

For the mean-square error, we also see from Figs. 7.1, 7.2, 7.3, and 7.4 that the performance of CKAPA compares well with that of CKLMS. In all cases, CKAPA provides faster convergence than CKLMS. This is due to the improved tracking ability of APA compared to LMS. Consider that if the original nonlinear problem is not recast to a completely linear problem in the transform space, the channel model used here, although stationary, may appear as non-stationary. For this scenario, APA-based algorithms may provide a benefit.

Finally, the performance of the kernels other than Gaussian, in particular the Bergman kernel, appears to be non-ideal based on the results obtained. This may be due to the choice of $\sigma$, $q$ parameters used. Although these parameters were obtained through extensive trial-and-error testing, the search was not exhaustive. The use of other kernels, both real and complex, may also be considered to improve performance.

## 7.4 Conclusion

We described here various forms of the complex kernel APA (CKAPA) algorithm. The new kernel-based methods improve convergence speed compared to CKLMS, and are suitable for real-time nonlinear applications using complex data.

The derivation of these algorithms uses Wirtinger's calculus for Hilbert spaces presented in [1] to obtain the gradient of the MSE cost function for APA, since the algorithm input is transform-mapped to either a complex or complexified RKHS space. Simulation results with various real and complex kernels show a faster convergence rate and improved performance of CKAPA over CKLMS algorithms.

## References


1. P. Bouboulis, S. Theodoridis, Extension of Wirtinger's calculus to reproducing Kernel Hilbert spaces and the complex Kernel LMS. IEEE Trans. Signal Process. **59**(3) (2011)
2. P. Bouboulis, S. Theodoridis, The complex Gaussian Kernel LMS algorithm, in *Proc. of Int. Conf. on Artificial Neural Networks* (2010)
3. P.S.R. Diniz, *Adaptive Filtering: Algorithms and Practical Implementation*, 3rd edn. (Springer Publishers, 2008)
4. W. Liu, J. Principe, S. Haykin, *Kernel Adaptive Filtering: A Comprehensive Introduction* (Wiley Publishers, 2010)
5. S. Haykin, *Adaptive Signal Processing*, 4th edn. (Prentice-Hall, 2002)

# Chapter 8
# Extending Complex Kernel Filters with Widely Linear Estimation

This chapter describes an approach for extending complex kernel-based adaptive algorithms (specifically CKLMS, CKAPA) based on the widely linear (WL) estimation approach of Chap. 3. A suitable estimator form using WL estimation is described, which is shown to simplify to a convenient form for implementation. An analysis of the potential performance benefit for the WL approach is provided (based on the polynomial kernel), and simulations are shown verifying the improvement.

## 8.1 Derivation of Algorithms

Considering the widely linear (WL) estimation approach (see Chap. 3), for combining WL estimation with CKLMS and CKAPA, we replace the previous estimator:

$$\hat{d}(n) = < \boldsymbol{\Phi}(\mathbf{z}(n)), \mathbf{w} >_S \tag{8.1}$$

with estimates

$$\hat{d}_{WL}(n) = < \boldsymbol{\Phi}(\mathbf{z}_{WL}(n)), \mathbf{w}_{WL} >_S \tag{8.2}$$

where

$$\mathbf{z}_{WL}(n) = \begin{bmatrix} \mathbf{z}(n) \\ \mathbf{z}^*(n) \end{bmatrix}$$

and $\mathbf{w}_{WL}$ are the optimum weights for the widely linear approach. For convenience of analysis (shown later), we define the function, $\xi(.)$:



T. K. Paul and T. Ogunfunmi, *Adaptive Filter Algorithms For Complex and Quaternion Data*, https://doi.org/10.1007/978-3-032-23798-9_8

$$\xi(\mathbf{z}(n)) = \mathbf{z}_{WL}(n) = \begin{bmatrix} \mathbf{z}(n) \\ \mathbf{z}^*(n) \end{bmatrix} \tag{8.3}$$

which will be used instead of the notation $\mathbf{z}_{WL}(n)$.

To evaluate the weights for estimator (8.2) using an LMS-based approach, we use the following weight update equation:

$$\mathbf{w}_{WL}(n) = \mathbf{w}_{WL}(n-1) + \mu e^*(n)\boldsymbol{\Phi}(\xi(\mathbf{z}(n))) \tag{8.4}$$

where $e(n) = d(n) - \hat{d}_{WL}(n)$ is the estimation error.

Assuming initial weights, $\mathbf{w}_{WL}(0) = 0$, at iteration $n$ we get

$$\mathbf{w}_{WL}(n) = \mu \sum_{k=1}^{n} e^*(k)\boldsymbol{\Phi}(\xi(\mathbf{z}(k))). \tag{8.5}$$

And the filter output is

$$\begin{aligned} \hat{d}(n) &=< \boldsymbol{\Phi}(\xi(\mathbf{z}(n))), \mathbf{w}_{WL}(n-1) >_S \\ &= \mu \sum_{k=1}^{n-1} e(k) < \boldsymbol{\Phi}(\xi(\mathbf{z}(n))), \boldsymbol{\Phi}(\xi(\mathbf{z}(k))) >_S \end{aligned} \tag{8.6}$$

Equation (8.6) forms the WL-CKLMS algorithm.

In a similar way, considering Chap. 7, an APA-based method (WL-CKAPA) can be seen to have the following form. Based on the $K$ most recent input vectors and observations:

$$\begin{aligned} &\mathbf{X}_{\xi,ap}(n) = [\boldsymbol{\Phi}(\xi(\mathbf{z}(n-K+1)))\ \boldsymbol{\Phi}(\xi(\mathbf{z}(n-K+2)))\ \ldots\ \boldsymbol{\Phi}(\xi(\mathbf{z}(n)))] \\ &\mathbf{d}_{ap}(n) = [d(n-K+1)\ d(n-K+2)\ \ldots\ d(n)]^T \end{aligned}$$

The weight update for WL-CKAPA may be expressed as

$$\mathbf{w}_{WL}(n) = \mathbf{w}_{WL}(n-1) + \mu\mathbf{X}_{\xi,ap}(n)\left(\mathbf{X}_{\xi,ap}^H(n)\mathbf{X}_{\xi,ap}(n) + \varepsilon\mathbf{I}\right)^{-1}\mathbf{e}_{ap}^*(n) \tag{8.7}$$

where $\mathbf{e}_{ap}(n) = \mathbf{d}_{ap}(n) - \hat{\mathbf{d}}_{WL,ap}(n)$ is the estimation error vector.

The weights at iteration $n$, assuming $\mathrm{w}(0) = 0$, can be shown to be of the following form:

$$\mathbf{w}_{WL}(n-1) = \sum_{j=1}^{n-1} a_j(n-1)\boldsymbol{\Phi}(\xi(\mathbf{z}(j))) \tag{8.8}$$

where $a_j(n-1)$, for $j = 1\ldots(n-1)$ are determined based on (8.7), yielding

$$a_k(i) = \begin{cases} \mu \tilde{\mathbf{e}}_K(i), & k = i \\ a_k(i-1) + \mu \tilde{\mathbf{e}}_{K+k-i}(i), & i-K+1 \leq k \leq i-1 \\ a_k(i-1), & 1 \leq k < i-K+1 \end{cases} \quad (8.9)$$

where $\tilde{\mathbf{e}}_k(n)$ is the kth element of vector $\tilde{\mathbf{e}}(i) = \left(\mathbf{X}^H_{\xi,ap}(n)\mathbf{X}_{\xi,ap}(n) + \varepsilon\mathbf{I}\right)^{-1} \mathbf{e}^*_{ap}(i)$.

The $a_j(n)$ coefficients are found with the same process as in Chap. 7.

The filter output for WL-CKAPA is

$$\hat{\mathbf{d}}_{WL,ap}(n) = \mathbf{X}^T_{\xi,ap}(n)\mathbf{w}^*_{WL}(n-1)$$
$$= \begin{bmatrix} \sum_{j=1}^{n-1} a^*_j(n-1) < \boldsymbol{\Phi}(\xi(\mathbf{z}(n-K+1))), \boldsymbol{\Phi}(\xi(\mathbf{z}(j))) >_S \\ \vdots \\ \sum_{j=1}^{n-1} a^*_j(n-1) < \boldsymbol{\Phi}(\xi(\mathbf{z}(n-1))), \boldsymbol{\Phi}(\xi(\mathbf{z}(j))) >_S \\ \sum_{j=1}^{n-1} a^*_j(n-1) < \boldsymbol{\Phi}(\xi(\mathbf{z}(n))), \boldsymbol{\Phi}(\xi(\mathbf{z}(j))) >_S \end{bmatrix} \quad (8.10)$$

where $\hat{\mathbf{d}}_{WL,ap}(n)$ is a vector of the $K$ most recent channel output estimates.

## 8.2 Widely Linear Kernels

For both methods, it can be observed that the difference between the original algorithms and the WL estimation-based algorithms is only in the kernel evaluation, which changes from

$$\kappa(\mathbf{u}, \mathbf{v}) = < \boldsymbol{\Phi}(\mathbf{u}), \boldsymbol{\Phi}(\mathbf{v}) >_S$$

to the form

$$\kappa_{WL}(\mathbf{u}, \mathbf{v}) = < \boldsymbol{\Phi}(\xi(\mathbf{u})), \boldsymbol{\Phi}(\xi(\mathbf{v})) >_S . \quad (8.11)$$

This is straightforward to see, considering the input vector transformation (8.3).

In order to convert the CKLMS, CKAPA algorithms (or any other complex kernel-based methods) to consider widely linear estimation, we may simply replace the complex kernel evaluation with a form considering WL estimation.

Here we illustrate with a few examples. Considering the standard Hermitian inner product, i.e.:

$$< \boldsymbol{\Phi}(\mathbf{u}), \boldsymbol{\Phi}(\mathbf{v}) >_S \; = \; \mathbf{v}^H \mathbf{u} \tag{8.12}$$

The WL estimation-based form is

$$\begin{aligned} < \boldsymbol{\Phi}(\xi(\mathbf{u})), \boldsymbol{\Phi}(\xi(\mathbf{v})) >_S \; &= \begin{bmatrix} \mathbf{v} \\ \mathbf{v}^* \end{bmatrix}^H \begin{bmatrix} \mathbf{u} \\ \mathbf{u}^* \end{bmatrix} \\ &= [\mathbf{v}^H \; \mathbf{v}^T] \begin{bmatrix} \mathbf{u} \\ \mathbf{u}^* \end{bmatrix} \\ &= \mathbf{v}^H \mathbf{u} + \mathbf{v}^T \mathbf{u}^* \\ &= 2\mathrm{Re}\{\mathbf{v}^H \mathbf{u}\} \end{aligned} \tag{8.13}$$

Use of Eq. (8.12) for (8.1) can be seen to reduce the CKLMS to an algorithm equivalent to the Complex LMS. This is because the kernel function used is the standard Hermitian inner product. Meanwhile, use of (8.13) for (8.2) reduces WL-CKLMS to the Widely Linear LMS of [1, 2], although both algorithms are now cast entirely in terms of inner products. A similar claim can be made considering (8.12) in the CKAPA, and (8.13) in the WL-CKAPA. Specifically, they become the Complex APA and the WL form of Complex APA, respectively.

We can also similarly consider other complex kernels. For example, with the complex Gaussian kernel:

$$< \boldsymbol{\Phi}(\mathbf{u}), \boldsymbol{\Phi}(\mathbf{v}) >_S \; = \; \exp\left( - \frac{\sum_{i=1}^{d} (u_i - v_i^*)^2}{\sigma^2} \right) \tag{8.14}$$

The WL estimation-based form is

$$\begin{aligned} &< \boldsymbol{\Phi}(\xi(\mathbf{u})), \boldsymbol{\Phi}(\xi(\mathbf{v})) >_S \\ &\quad = \exp\left( - \frac{\sum_{i=1}^{d} (u_i - v_i^*)^2 + \sum_{i=1}^{d} (u_i^* - v_i)^2}{\sigma^2} \right) \\ &\quad = \exp\left( - \frac{\sum_{i=1}^{d} 2\mathrm{Re}\{(u_i - v_i^*)^2\}}{\sigma^2} \right) \end{aligned} \tag{8.15}$$

The kernel function (8.15) represents the widely linear version of the complex Gaussian kernel (8.14). We will also consider the polynomial kernel later in this section.

An important consideration, however, is whether the function $\kappa_{WL}(\cdot,\cdot)$ of (8.11) remains a kernel function when $\kappa(\cdot,\cdot)$ is a kernel. To determine this, we consider some properties of kernel functions described in Paulson [3].

To determine if function $\kappa_{WL}(\cdot,\cdot)$ of (8.11) is a kernel function if $\kappa(\cdot,\cdot)$ is a kernel, we consider that the inputs are transformed by $\xi(\cdot)$ before applying $\kappa(\cdot,\cdot)$ in (8.11). This scenario can be described as follows. With the original kernel function $\kappa : X \times X \rightarrow$

$\mathbb{C}$, and some function $\xi : S \to X$, we can then define the composite function as $\kappa \circ \xi : S \times S \to \mathbb{C}$, which denotes the operation: $(\kappa \circ \xi)\ (\mathbf{x}_1, \mathbf{x}_2) = \kappa(\xi(\mathbf{x}_1), \xi(\mathbf{x}_2))$.

From [3], for any kernel $\kappa$, the composite function $\kappa \circ \xi$ is a kernel function on $S$ for any arbitrary function, $\xi$. The associated Hilbert space $\mathcal{H}(\kappa \circ \xi)$ is referred to as the pull-back of $\mathcal{H}(\kappa)$ along $\xi$, and can be written as: $\mathcal{H}(\kappa \circ \xi) = \{f \circ \xi\ :\ f\ \in \mathcal{H}(\kappa)\}$. Note that the function $\kappa_{WL}(\cdot,\cdot)$ of (8.11) can always be considered a kernel when $\kappa(\cdot,\cdot)$ is a kernel.

## 8.3 Benefit for Complex Nonlinear Channels

To understand the benefits of the widely linear methods described for learning with complex nonlinear channels, we consider an example using the polynomial kernel.

The polynomial kernel may be written as [4]

$$K(\mathbf{u}, \mathbf{v}) = (1 + \mathbf{v}^T\mathbf{u})^q$$

Using the Hermitian inner product, assuming $\mathbf{u}$, $\mathbf{v}$ are complex:

$$K(\mathbf{u}, \mathbf{v}) = (1 + \mathbf{v}^H\mathbf{u})^q. \tag{8.16}$$

Note that we have not yet considered the use of WL estimation.

To simplify the analysis, we consider the $q = 2$ case only. The results can be seen to extend for larger values of $q$. Also, we will consider $\mathbf{u}$, $\mathbf{v}$ as vectors of length $M = 2$ (again to simply the analysis). The length $M$ of $\mathbf{u}$, $\mathbf{v}$ corresponds to the input vector length used in estimating the nonlinear channel (same as $\mathbf{z}(n)$ in (8.1)). The effect of this length will be considered in the analysis to follow.

For $q = 2$, Eq. (8.16) can be expanded as

$$\begin{aligned} K(\mathbf{u}, \mathbf{v}) &= (1 + \mathbf{v}^H\mathbf{u})^2 \\ &= 1 + 2\mathbf{v}^H\mathbf{u} + (\mathbf{v}^H\mathbf{u})^2 \\ &= 1 + 2(v_1^* u_1 + v_2^* u_2) + (v_1^* u_1 + v_2^* u_2)^2 \\ &= 1 + 2v_1^* u_1 + 2v_2^* u_2 + v_1^{*2} u_1^2 + 2v_1^* v_2^* u_1 u_2 + v_2^{*2} u_2^2 \end{aligned} \tag{8.17}$$

which can be expressed using the form

$$K(\mathbf{u}, \mathbf{v}) = \varphi(\mathbf{v})^H \varphi(\mathbf{u}) \tag{8.18}$$

where:

$$\varphi(\mathbf{u}) = \begin{bmatrix} 1 & \sqrt{2}\,u_1 & \sqrt{2}\,u_2 & u_1^2 & \sqrt{2}\,u_1 u_2 & u_2^2 \end{bmatrix}^T \tag{8.19}$$

$$\varphi(\mathbf{v}) = \left[1 \ \sqrt{2}\,v_1 \ \sqrt{2}\,v_2 \ v_1^2 \ \sqrt{2}\,v_1v_2 \ v_2^2\right]^T.$$

Note that the function (8.16) can be equivalently written in the form $\varphi(\mathbf{v})^H\varphi(\mathbf{u})$, and therefore satisfies the Mercer condition for kernel functions (see Chap. 2, also [4]). Also note that the complex polynomial kernel form of (8.16) can be seen as defining the inner product for a nonlinear transform space (i.e. complex RKHS of the complex kernel), where $\varphi(\cdot)$ is the transform mapping function.

Considering (8.18), the 2-dimensional input vector is nonlinearly mapped to a transform space with dimension size 6. The transform space can be considered as follows. Based on Eq. (8.4), we can see function $\varphi(\cdot)$ of (8.19) may be viewed as the definition of mapping $\boldsymbol{\Phi}(\cdot)$ when (8.16) is the kernel being considered.

If we evaluate $\boldsymbol{\Phi}(\mathbf{z}(n))$ (when $M = 2$):

$$\begin{aligned}\boldsymbol{\Phi}(\mathbf{z}(n)) &= \varphi(\mathbf{z}(n)) \\ &= \left[1 \ \sqrt{2}\,z(n) \ \sqrt{2}\,z(n-1) \ z^2(n) \ \sqrt{2}\,z(n)z(n-1) \ z^2(n-1)\right]^T.\end{aligned} \tag{8.20}$$

The input is transformed to a space where all the various correlations of the elements of $\mathbf{z}(n)$, up to order $q = 2$, are the dimensions of the transform space for mapping $\boldsymbol{\Phi}(\cdot)$ (i.e. dimensions are 1, $z(n)$, $z(n-1)$, $z^2(n)$, $z(n)\cdot z(n-1)$, $z^2(n-1)$). Therefore, linear weights, $\mathbf{w}$, applied to transformed input, $\boldsymbol{\Phi}(\mathbf{z}(n))$, can learn nonlinear channels up to order $q = 2$.

However, to improve the nonlinear MS estimation performance with complex data, we should use an approach that allows the moments of estimate $\hat{d}(n)$ to be described by moments of order $k$ of inputs $z(n)$, $z(n-1)$ and their conjugates, similar to the widely linear estimator of Chap. 3. This clearly is not the case using (8.19), since the conjugate forms of $z(n)$, $z(n-1)$ are not present.

Now, we consider the widely linear form of the kernel of (8.16):

$$\begin{aligned}K(\xi(\mathbf{u}), \xi(\mathbf{v})) &= \left(1 + \begin{bmatrix}\mathbf{v}\\ \mathbf{v}^*\end{bmatrix}^H \begin{bmatrix}\mathbf{u}\\ \mathbf{u}^*\end{bmatrix}\right)^q \\ &= \left(1 + \mathbf{v}^H\mathbf{u} + \mathbf{v}^T\mathbf{u}^*\right)^q \\ &= \left(1 + 2\mathrm{Re}\{\mathbf{v}^H\mathbf{u}\}\right)^q.\end{aligned} \tag{8.21}$$

Considering the $q = 2$, $M = 2$ case with Eq. (8.21):

$$\begin{aligned}K(\mathbf{u}, \mathbf{v}) &= (1 + \mathbf{v}^H\mathbf{u} + (\mathbf{v}^H\mathbf{u})^*)^2 \\ &= 1 + 2((\mathbf{v}^H\mathbf{u}) + (\mathbf{v}^H\mathbf{u})^*) + ((\mathbf{v}^H\mathbf{u}) + (\mathbf{v}^H\mathbf{u})^*)^2 \\ &= 1 + 2(v_1^*u_1 + v_2^*u_2 + v_1u_1^* + v_2u_2^*) + (v_1^*u_1 + v_2^*u_2 + v_1u_1^* + v_2u_2^*)^2 \\ &= 1 + 2v_1^*u_1 + 2v_2^*u_2 + 2v_1u_1^* + 2v_2u_2^*\end{aligned}$$

$$
\begin{aligned}
&+ v_1^{*2}u_1^2 + v_1^*v_2^*u_1u_2 + v_1^*v_1u_1u_1^* + v_1^*v_2u_1u_2^* + v_1^*v_2^*u_1u_2 \\
&+ v_2^{*2}u_2^2 + v_2^*v_1u_2u_1^* + v_2^*v_2u_2u_2^* \\
&+ v_1^*v_1u_1u_1^* + v_2^*v_1u_2u_1^* + v_1^2u_1^{*2} + v_1v_2u_1^*u_2^* \\
&+ v_1^*v_2u_1u_2^* + v_2^*v_2u_2u_2^* + v_1v_2u_1^*u_2^* + v_2^2u_2^{*2} \\
= &\; 1 + 2v_1^*u_1 + 2v_2^*u_2 + 2v_1u_1^* + 2v_2u_2^* \\
&+ 2v_1^*v_2^*u_1u_2 + 2v_1^*v_1u_1u_1^* + 2v_1^*v_2u_1u_2^* + 2v_2^*v_1u_2u_1^* \\
&+ 2v_2^*v_2u_2u_2^* + 2v_1v_2u_1^*u_2^* + v_1^{*2}u_1^2 + v_2^{*2}u_2^2 + v_1^2u_1^{*2} + v_2^2u_2^{*2} \qquad (8.22)
\end{aligned}
$$

which can be written in the form

$$
K(\xi(\mathbf{u}), \xi(\mathbf{v})) = \varphi_\xi(\mathbf{v})^H \varphi_\xi(\mathbf{u}) \qquad (8.23)
$$

where:

$$
\begin{aligned}
\varphi_\xi(\mathbf{v}) =& [1 \;\; \sqrt{2}\cdot v_1 \;\; \sqrt{2}\cdot v_2 \;\; \sqrt{2}\cdot v_1^* \;\; \sqrt{2}\cdot v_2^* \;\; \sqrt{2}\cdot v_1v_2 \;\; \sqrt{2}\cdot v_1v_1^* \;\; \sqrt{2}\cdot v_1v_2^* \\
& \sqrt{2}\cdot v_2v_1^* \;\; \sqrt{2}\cdot v_2v_2^* \;\; \sqrt{2}\cdot v_1^*v_2^* \;\; v_1^2 \;\; v_2^2 \;\; v_1^{*2} \;\; v_2^{*2}\,]^T \qquad (8.24)
\end{aligned}
$$

$$
\begin{aligned}
\varphi_\xi(\mathbf{u}) =& [1 \;\; \sqrt{2}\cdot u_1 \;\; \sqrt{2}\cdot u_2 \;\; \sqrt{2}\cdot u_1^* \;\; \sqrt{2}\cdot u_2^* \;\; \sqrt{2}\cdot u_1u_2 \;\; \sqrt{2}\cdot u_1u_1^* \;\; \sqrt{2}\cdot u_1u_2^* \\
& \sqrt{2}\cdot u_2u_1^* \;\; \sqrt{2}\cdot u_2u_2^* \;\; \sqrt{2}\cdot u_1^*u_2^* \;\; u_1^2 \;\; u_2^2 \;\; u_1^{*2} \;\; u_2^{*2}\,]^T.
\end{aligned}
$$

Using $\varphi_\xi(\cdot)$ of Eq. (8.24), the 2-dimensional input vector is nonlinearly mapped to a transform space of dimension size 15.

Considering the transform space of $\varphi_\xi(\cdot)$ in a similar fashion to the transform space of $\varphi(\cdot)$, the dimensions of the transform space for mapping $\mathbf{\Phi}(\xi(\mathbf{z}(n)))$ can be seen to consist of

$$
\begin{aligned}
&1, z(n), z(n-1), z^*(n), z^*(n-1), z(n)\cdot z(n-1), z(n)\cdot z^*(n), z(n)\cdot z^*(n-1), \\
&z(n-1)\cdot z^*(n), z(n-1)\cdot z^*(n-1), z^*(n)\cdot z^*(n-1), z^2(n), z^2(n-1), \\
&z^{*2}(n), z^{*2}(n-1)
\end{aligned}
$$

The input is now transformed into a space where all the various correlations of the elements of $\mathbf{z}(n)$ along with their conjugates, up to order $q = 2$, form the dimensions of the transform space for the mapping.

Use of kernel (8.22) (i.e. transform mapping (8.24)) should allow for improved nonlinear MS estimation performance with complex data, because the moments of the estimate can be completely described using moments of order $k$ of the input samples, $z(n)$, $z(n-1)$ and their conjugates. The result obtained can also be seen to extend for larger values of $q$ and $M$.

The use of widely linear estimation can be seen to provide benefit when using the polynomial kernel (8.16) and the standard Hermitian inner product (8.12). A proof of the benefits for the complex Gaussian kernel (8.14) may be similarly considered based on the RKHS description of [5] and Appendix A. Simulations are used to verify the benefit.

## 8.4 Simulation Results

Here, we compare the performance of the WL-CKLMS and WL-CKAPA with their CKLMS, CKAPA counterparts, for a nonlinear channel estimation task. The testing is performed using the complex Gaussian kernel and circular, noncircular complex data.

The nonlinear channel model used consisted of the widely linear filter:

$$\begin{aligned} t(n) = &(-0.9 + 0.8i) \cdot s(n) + (0.6 - 0.7i) \cdot s(n-1) \\ &+ (0.4 + 0.7i) \cdot s^*(n) + (-0.3 - 0.6i) \cdot s^*(n-1) \end{aligned} \tag{8.25}$$

followed by memoryless nonlinearity:

$$q(n) = t(n) + (0.1 - 0.15i) \cdot t^2(n) + (0.06 + 0.05i) \cdot t^3(n) \tag{8.26}$$

The approach is based on a Weiner nonlinear model [6], though modified considering widely linear estimation. The signal $r(n)$ at the receiver is corrupted using white Gaussian noise, i.e. $r(n) = q(n) + v(n)$. The noise, $v(n)$, is added such that the SNR is 16 dB at the receiver. The complex input data was generated using

$$s(n) = 0.7(\sqrt{1-\rho^2}X(n) + i \cdot \rho Y(n))$$

where $X(n)$, $Y(n)$ are Gaussian random variables with unit variance, and $\rho$ is a variable controlling circularity. The value $\rho = 0.7071$ generates circular input, while values close to either 0 or 1 generate highly noncircular input. Here we use $\rho = 0.7071$ for tests with circular input, and $\rho = 0.1$ for tests with noncircular input.

The following tests are shown here:

Fig. 8.1: WL-CKLMS, CKLMS with Circular Input

Fig. 8.2: WL-CKLMS, CKLMS with Noncircular Input

Fig. 8.3: WL-CKAPA, CKAPA with Circular Input

Fig. 8.4: WL-CKAPA, CKAPA with Noncircular Input

Each of the tests includes simulation results with the corresponding algorithm based on the complexified form of the real Gaussian kernel described in reference [7]. Also, $K = 3$ was used for WL-CKAPA and CKAPA.

For both Figs. 8.1 and 8.2, we see WL-CKLMS improved the MSE performance compared to CKLMS (a 15 dB difference for circular input, 3–4 dB for noncircular input). The substantial difference for the circular input case is similar to the results of [2] for the same scenario (i.e. use of circular data for a channel with a nonlinear

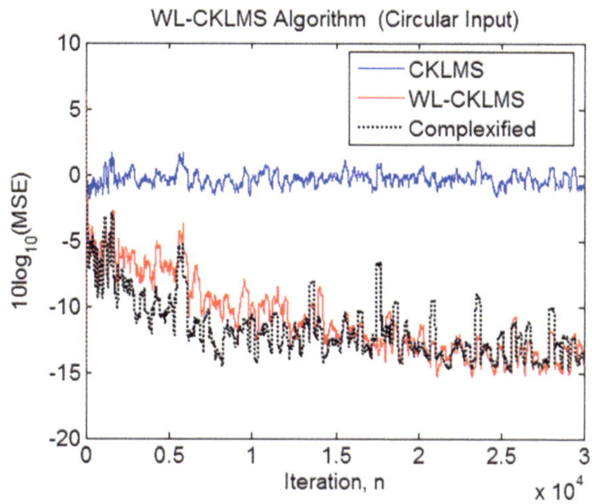


**Fig. 8.1** WL-CKLMS, CKLMS with circular input

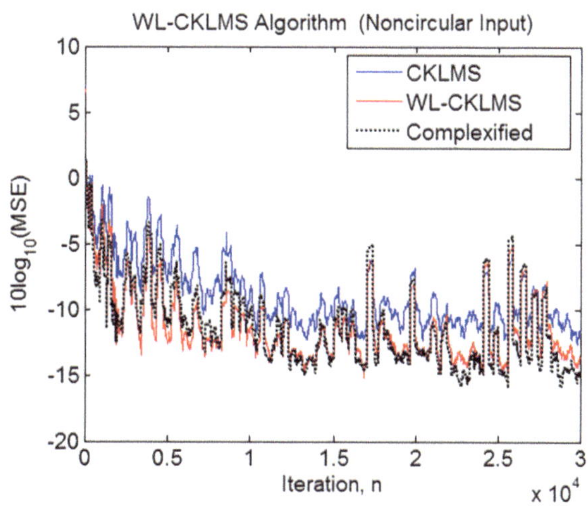


**Fig. 8.2** WL-CKLMS, CKLMS with noncircular input

form modeled well by WL estimation), and a detailed reason for the difference in MSE is described therein.

However, MSE performance for the complexified kernel method is essentially similar to WL-CKLMS. Both methods appear to work well with both circular data, as well as noncircular data. Any further improvement may be achieved by adjusting the kernel parameter $\sigma$ of (8.14), which affects the nonlinear mapping. This parameter was chosen in both cases through extensive trial-and-error testing, and the performance for both may vary.

For Figs. 8.3 and 8.4, we see similar test results based on the WL-CKAPA and CKAPA. The MSE performance difference between WL-CKAPA and CKAPA in steady-state again differs by 15 dB for circular data and 3–4 dB for noncircular data,

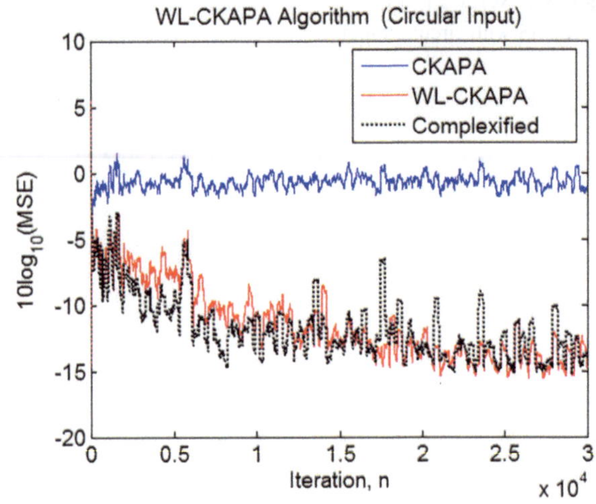


**Fig. 8.3** WL-CKAPA, CKAPA with Circular Input

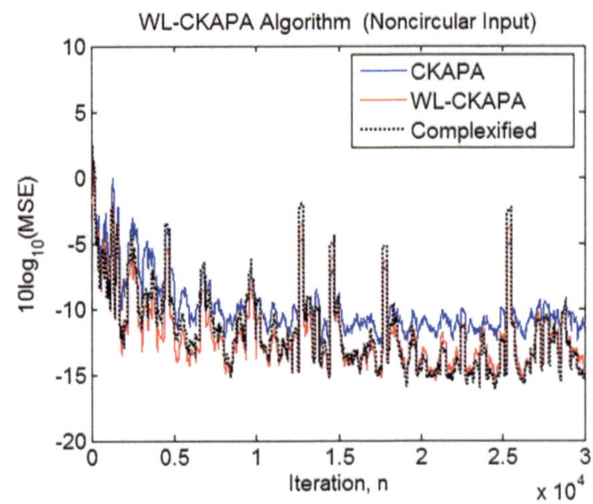


**Fig. 8.4** WL-CKAPA, CKAPA with Noncircular Input

respectively. And the MSE performance of WL-CKAPA compared to the complexified approach is similar, with suboptimalities likely due to the choice of kernel parameter $\sigma$, similar to before.

However, the convergence benefit for the APA version compared to LMS is not clearly visible from Fig. 8.1, 8.2, 8.3, and 8.4. APA-based algorithms are known to provide a convergence benefit, however, typically for colored input signals [8, 9].

A simulation result illustrating the convergence rate benefit for WL-CKAPA compared to WL-CKLMS can be seen in Fig. 8.5, where a nonlinear equalization test is performed (instead of channel estimation) using the same nonlinear channel used for Figs. 8.1, 8.2, 8.3, and 8.4 (i.e. Eqs. (8.25) and (8.26)). The system model is illustrated in Fig. 8.6. The benefit of faster initial convergence is demonstrated here

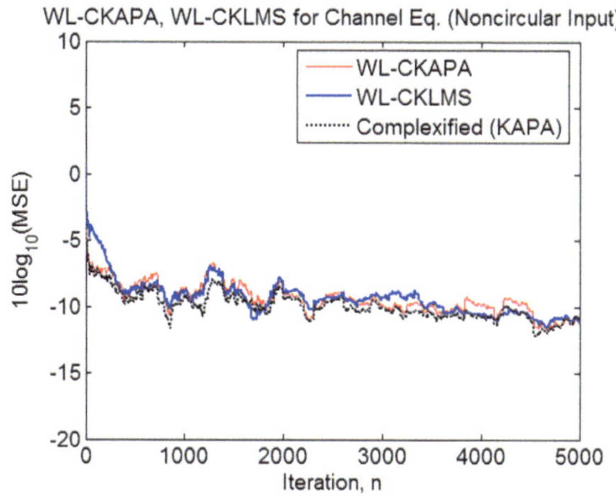


**Fig. 8.5** WL-CKAPA, WL-CKLMS with colored noncircular input

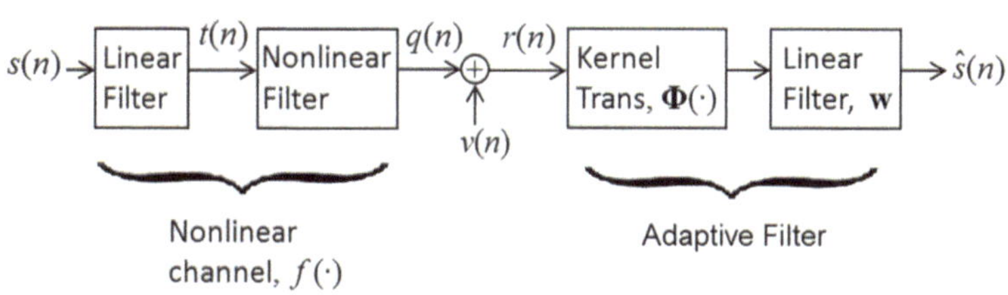


**Fig. 8.6** System model for channel equalization (used for Fig. 8.5)

for WL-CKAPA. However, channel equalization can often lead to enhancement of noise, affecting misadjustment performance.

Finally, we note that the complexified kernel method has similar performance to the WL estimation-based methods described here. However, here we described an approach for improving the performance of the complex kernel-based methods from [7, 10], allowing for improved MSE performance, as well as broadening the understanding of complex kernel functions.

## 8.5 Conclusion

Here, we described an extension of the CKLMS and CKAPA complex kernel methods based on widely linear estimation. The approach used involved deriving widely linear versions of complex kernel functions, and may easily be applied to other complex kernel methods, aside from CKLMS and CKAPA.

Also, an analysis of the estimation benefit based on widely linear estimation (considering the polynomial kernel) was illustrated, indicating that the full statistical information needed for MS estimation with complex data is available using this approach. Simulation results demonstrated the benefits compared to the original complex kernel methods, and similarity with the complexified method from [7].

## References


1. B. Picinbono, Bernard and Chevalier, Pascal, "widely linear estimation with complex data. IEEE Trans. Signal Process. **43**(8), 2030–2033 (1995)
2. T. Adali, H. Li, R. Aloyious, On properties of the widely linear MSE filter and its LMS implementation", in *Proc. Conf. Inf. Sci. Syst. (CISS)* (Baltimore, MD, 2009)
3. V.I. Paulsen, An Introduction to the Theory of Reproducing Kernel Hilbert Spaces. http://www.math.uh.edu/vern/rkhs.pdf
4. W. Liu, J. Principe, S. Haykin, *Kernel Adaptive Filtering: A Comprehensive Introduction* (Wiley Publishers, 2010)
5. I. Steinwart, D. Hush, C. Scovel, An explicit description of the reproducing Kernel Hilbert spaces of Gaussian RBF kernels. IEEE Trans. Inf. Theory **52**(10) (2006)
6. T. Ogunfunmi, *Adaptive Nonlinear System Identification: The Volterra and Wiener Model Approaches* (Springer Publishers, 2007)
7. P. Bouboulis, S. Theodoridis, Extension of Wirtinger's calculus to reproducing Kernel Hilbert spaces and the complex Kernel LMS. IEEE Trans. Signal Process. **59**(3) (2011)
8. S. Haykin, *Adaptive Signal Processing*, 4th edn. (Prentice-Hall, 2002)
9. P.S.R. Diniz, *Adaptive Filtering: Algorithms and Practical Implementation*, 3rd edn. (Springer Publishers, 2008)
10. P. Bouboulis, S. Theodoridis, The complex Gaussian Kernel LMS algorithm, in *Proc. of Int. Conf. on Artificial Neural Networks* (2010)

# Chapter 9
# Convergence Analysis of Complex Kernel LMS Algorithm

This chapter presents an analysis of convergence of the NCKLMS algorithm of Chap. 4 (simply referred to as CKLMS here). The approach follows the analysis by Parreira et al. [1, 2] for KLMS. The analysis is performed for both the complex Gaussian kernel, as well as the complexified real Gaussian kernel (or complexified Gaussian kernel for short). A complex Gaussian sequence containing a parameter for controlling circularity was used as input. The resulting expressions allow theoretical MSE predictions based on the circularity of the complex input, in addition to its autocorrelation.

## 9.1 Analysis Preliminaries

For the analysis of complex KLMS, with the algorithm as described in Chap. 4, we assume complex input of the form:

$$\boldsymbol{u}(n) = \sqrt{1-\rho^2} \cdot \boldsymbol{x}_{re}(n) + i\rho \boldsymbol{x}_{im}(n) \tag{9.1}$$

where both $\{\mathbf{x}_{re}(n)\}$ and $\{\mathbf{x}_{im}(n)\}$ are sequences of real i.i.d. Gaussian vectors with possibly correlated components, and where each component of $\mathbf{x}_{re}(n)$ and $\mathbf{x}_{im}(n)$ is zero-mean, with variance $\sigma_{\mathbf{x}}{}^2$. Sequences $\{\mathbf{x}_{re}(n)\}$ and $\{\mathbf{x}_{im}(n)\}$ are assumed to be independent.

The vector $\{\mathbf{u}(n)\}$ is a sequence of complex i.i.d Gaussian vectors with possibly correlated components. Each of the components of vector $\mathbf{u}(n)$ has a circularity controlled by parameter $\rho$, where $\rho = 1/\sqrt{2}$ results in each component of $\mathbf{u}(n)$ having a circular distribution, and where $\rho$ close to 0 or 1 results in each component having a highly non-circular distribution.

For convenience of notation, we will use the form



T. K. Paul and T. Ogunfunmi, *Adaptive Filter Algorithms For Complex and Quaternion Data*, https://doi.org/10.1007/978-3-032-23798-9_9

$$\boldsymbol{u}(n) = \rho_{re}\boldsymbol{x}_{re}(n) + i\rho_{im}\boldsymbol{x}_{im}(n) \tag{9.2}$$

where $\rho_{re} = \sqrt{1-\rho^2}$, $\rho_{im} = \rho$.

In addition, we also consider the following kernels [3, 4]:

Complex Gaussian Kernel:

$$< \boldsymbol{\Phi}(\boldsymbol{u}(m)), \boldsymbol{\Phi}(\boldsymbol{u}(n)) >_{\mathscr{H}} = \exp\left[\frac{-1}{2\xi^2}\sum_{k=1}^{L}\left[u_k(m) - u_k^*(n)\right]^2\right] \tag{9.3}$$

Complexified Gaussian Kernel:

$$\begin{aligned} &< \boldsymbol{\Phi}(\boldsymbol{u}(m)), \boldsymbol{\Phi}(\boldsymbol{u}(n)) >_{\mathscr{H}} \\ &= 2\exp\left[\frac{-1}{2\xi^2}\left(\sum_{k=1}^{L}\mathrm{Re}\{u_k(m) - u_k(n)\}^2 + \sum_{k=1}^{L}Im\{u_k(m) - u_k(n)\}^2\right)\right] \end{aligned} \tag{9.4}$$

where $u_k\ (n)$ is the $k$th element of $\mathbf{u}(n)$, and $L$ is the vector length.

For the analysis, we use the normalized CKLMS version from Chap. 4, and evaluate

$$\kappa_{norm}(\boldsymbol{u}(m), \boldsymbol{u}(n)) = \frac{< \boldsymbol{\Phi}(\boldsymbol{u}(m)), \boldsymbol{\Phi}(\boldsymbol{u}(n)) >_{\mathscr{H}}}{< \boldsymbol{\Phi}(\boldsymbol{u}(m)), \boldsymbol{\Phi}(\boldsymbol{u}(m)) >_{\mathscr{H}}} \tag{9.5}$$

For the two kernel functions, the normalized forms are as follows:

Normalized Complex Gaussian Kernel:

$$\begin{aligned} &\kappa_{norm}(\boldsymbol{u}(m), \boldsymbol{u}(n)) \\ &= \exp\left[\frac{-1}{2\xi^2}\left(\sum_{k=1}^{L}\left[u_k(m) - u_k^*(n)\right]^2 - \sum_{k=1}^{L}\left[u_k(m) - u_k^*(m)\right]^2\right)\right] \end{aligned} \tag{9.6}$$

Normalized Complexified Gaussian Kernel:

$$\begin{aligned} &\kappa_{norm}(\boldsymbol{u}(m), \boldsymbol{u}(n)) \\ &= \exp\left[\frac{-1}{2\xi^2}\left(\sum_{k=1}^{L}\mathrm{Re}\{u_k(m) - u_k(n)\}^2 + \sum_{k=1}^{L}Im\{u_k(m) - u_k(n)\}^2\right)\right] \end{aligned} \tag{9.7}$$

For Eq. (9.7), note $\kappa_{norm}(\mathbf{u}(m), \mathbf{u}(n)) = \kappa_{norm}(\mathbf{u}(n), \mathbf{u}(m))$.

For notation convenience, we will refer to the normalized form $\kappa_{norm}(.,.)$ simply as $\kappa(.,.)$ in the following section, and consider the normalized CKLMS algorithm (4.25) with the form:

$$\hat{d}(n) = \eta \sum_{k=1}^{n-1} e(k) \cdot \kappa(\boldsymbol{u}(n), \boldsymbol{u}(k))$$

where $\kappa(\boldsymbol{u}(n), \boldsymbol{u}(k))$ is either Eq. (9.6) or (9.7).

## 9.2 Convergence of Complex Kernel LMS

We now consider the convergence behavior of complex KLMS. The methodology used follows [1] although differs in the complex case due to the autocorrelation and covariance matrix factoring. Both the complex and complexified kernels forms are considered. The analysis here allows the effect of input circularity on the performance of an algorithm defined on a complex RKHS to be studied.

For complex KLMS, the squared error can be expressed as

$$\sum_{n=1}^{N} |d(n) - \psi(\boldsymbol{u}(n))|^2 \tag{9.8}$$

where $d(n)$ is the desired response of the nonlinear system, $\psi(\cdot)$ is the estimate of the nonlinear channel (note $\psi(\mathbf{u}(n)) = \langle \kappa(\cdot, \boldsymbol{u}(n)), \psi(\cdot)\rangle_{\mathbb{H}})$), and $N$ is number of data points.

Based on the representer theorem [1, 5], the function $\psi(\cdot)$ minimizing (9.8) has the form

$$\psi(\cdot) = \sum_{n=1}^{N} \alpha_n \cdot \kappa(\cdot, \boldsymbol{u}(n)) \tag{9.9}$$

where $\alpha_1,\ldots,\alpha_N$ are complex-valued weights. Therefore the optimization problem involves finding $\alpha_1, \ldots, \alpha_N$, which may be stated as minimizing

$$\left\| \boldsymbol{d} - \boldsymbol{K}\boldsymbol{\alpha}^* \right\|^2 \tag{9.10}$$

where $\boldsymbol{\alpha} = [\alpha_1, \ldots, \alpha_N]^T$, $\mathbf{K}$ is the matrix defined $\mathbf{K}_{p,l} = \kappa(\mathbf{u}(l), \mathbf{u}(p))$, and $\mathbf{d} = [d(1) \ldots d(n)]^T$.

Since data points $N$ may grow without bound, a finite model may be used, based on only the $M$ vectors: $\kappa\left(\cdot, \boldsymbol{u}\left(\omega_j\right)\right)$ where $j = 1..M$, referred to as the dictionary.

Therefore Eq. (9.9) becomes

$$\psi(\cdot) = \sum_{j=1}^{M} \alpha_j \cdot \kappa\left(\cdot, \boldsymbol{u}\left(\omega_j\right)\right) \tag{9.11}$$

Many approaches exist for choosing $M$ dictionary vectors. We use the coherence-based sparsification rule. With this method, $\kappa(\cdot, \boldsymbol{u}(l))$ is added if it satisfies

$$\max_j \left|\kappa\left(\boldsymbol{u}(l), \boldsymbol{u}(\omega_j)\right)\right| \leq \varepsilon_0 \tag{9.12}$$

where $\varepsilon_0$ is a parameter determining the sparsification level. It was shown using this sparsification rule, even as the number of input vectors grows unbounded, the dictionary size $M$ remains bounded (i.e., finite) for real kernels [6]. This can be seen to apply for complexified kernels as well, considering how the complexified form extends a real kernel for complex input. Here we assume, with complex kernels also, $M$ remains finite. This may require verification, however, following [6].

For the analysis here we assume that dictionary size $M$ is a constant, and its dictionary vectors are statistically described using (9.1). To find the MSE of CKLMS, consider the CKLMS estimate at iteration $n$, which can be expressed as

$$\hat{d}(n) = \kappa\langle(\cdot, \boldsymbol{u}(n)), \psi(\cdot)\rangle_{\mathbb{H}} = \sum_{j=1}^{M} \alpha_j^* \cdot \kappa\left(\boldsymbol{u}(\omega_j), \boldsymbol{u}(n)\right) \tag{9.13}$$

which can be written as

$$\hat{d}(n) = \boldsymbol{\alpha}^H(n)\boldsymbol{\kappa}_\omega(n) \tag{9.14}$$

where $\alpha(n) = \left[\boldsymbol{\alpha}_1(n), \boldsymbol{\alpha}_2(n), \ldots., \boldsymbol{\alpha}_m(n)^T\right], \boldsymbol{\kappa}_\omega(n) = [\kappa(\boldsymbol{u}(\omega_1), \boldsymbol{u}(n)), \ldots, \kappa(\boldsymbol{u}(\omega_M), \boldsymbol{u}(n))]^T$.

With estimation error $e(n) = d(n) - \hat{d}(n)$, the MSE is

$$J_{ms}(n) = E\left[|d(n)|^2\right] - \boldsymbol{p}_{kd}^H\boldsymbol{\alpha}(n) - \boldsymbol{\alpha}^H(n)\boldsymbol{p}_{kd} + \boldsymbol{\alpha}^H(n)\boldsymbol{R}_{\kappa\kappa}\boldsymbol{\alpha}(n) \tag{9.15}$$

where $\mathbf{R}_{\mathbf{\kappa\kappa}} = E[\boldsymbol{\kappa}_\omega(n)\boldsymbol{\kappa}_\omega{}^H(n)]$ is the correlation matrix of transformed input $\boldsymbol{\kappa}_\omega(n)$, and $\mathbf{p}_{\mathbf{k}d} = E[d^*(n)\boldsymbol{\kappa}_\omega(n)]$ is the cross-correlation vector of $\boldsymbol{\kappa}_\omega(n)$ and $d(n)$.

With $\mathbf{R}_{\mathbf{\kappa\kappa}}$ positive definite, optimal weights for $\boldsymbol{\alpha}(n)$ are

$$\boldsymbol{\alpha}_{opt} = \boldsymbol{R}_{\kappa\kappa}^{-1}\boldsymbol{p}_{kd} \tag{9.16}$$

yielding minimum MSE

$$J_{\min} = E\left[|d(n)|^2\right] - \boldsymbol{p}_{kd}^H\boldsymbol{R}_{\kappa\kappa}^{-1}\boldsymbol{p}_{kd} \tag{9.17}$$

For the CKLMS algorithm, finding the LMS-based weight update considering estimator (9.14) yields

$$\boldsymbol{\alpha}(n+1) = \boldsymbol{\alpha}(n) + \eta e^*(n)\boldsymbol{\kappa}_\omega(n) \tag{9.18}$$

Defining weight error, $\boldsymbol{v}(n) = \boldsymbol{\alpha}(n) - \boldsymbol{\alpha}_{opt}$, we obtain the weight-error update:

$$\boldsymbol{v}(n+1) = \boldsymbol{v}(n) + \eta e^*(n)\boldsymbol{\kappa}_\omega(n) \tag{9.19}$$

The estimation error can be expressed

$$e(n) = d(n) - \boldsymbol{\kappa}_\omega^T(n)\boldsymbol{v}^*(n) - \boldsymbol{\kappa}_\omega^T(n)\boldsymbol{\alpha}_{opt}^*(n) \tag{9.20}$$

where the optimal estimation error (i.e., weight-error zero) is

$$e_0(n) = d(n) - \boldsymbol{\kappa}_\omega^T(n)\boldsymbol{\alpha}_{opt}^*(n) \tag{9.21}$$

For the mean weight-error behavior, combining (9.19) and (9.20):

$$\begin{aligned}\boldsymbol{v}(n+1) &= \boldsymbol{v}(n) + \eta d^*(n)\boldsymbol{\kappa}_\omega(n) \\ &\quad - \eta\boldsymbol{\kappa}_\omega^H(n)\boldsymbol{v}(n)\boldsymbol{\kappa}_\omega(n) - \eta\boldsymbol{\kappa}_\omega^H(n)\boldsymbol{\alpha}_{opt}(n)\boldsymbol{\kappa}_\omega(n)\end{aligned} \tag{9.22}$$

For tractability of analysis, we assume $\mathbf{\kappa}_\omega(n)\mathbf{\kappa}_\omega{}^H(n)$ and $\mathbf{v}(n)$ are independent, referred to as the modified independence assumption (**MIA**). The **MIA** is known to work well for adaptive filter analysis, and may be seen to be less strict compared to the classical independence assumption between $\mathbf{\kappa}_\omega(n)$ and $\mathbf{v}(n)$ [1].

Considering (9.20) and using the **MIA**, the MSE can be shown to be

$$J_{ms}(n) = J_{\min} + tr\{\boldsymbol{R}_{\kappa\kappa}\boldsymbol{C}_{\boldsymbol{v}}(n)\} \tag{9.23}$$

where $\mathbf{C}_{\mathbf{v}}(n) = E[\mathbf{v}(n)\mathbf{v}^H(n)]$ is the weight-error covariance matrix, and $J_{\min}$ is minimum MSE from before.

Finding the MSE behavior requires evaluating the matrices $\mathbf{R}_{\kappa\kappa}$ and $\mathbf{C}_{\mathbf{v}}(n)$. Here we evaluate these two matrices based on both the complex and complexified Gaussian kernel.

## 9.2.1 Finding Matrix $R_{\kappa\kappa}$

For correlation matrix $\mathbf{R}_{\kappa\kappa}$, the elements are

$$[\boldsymbol{R}_{\kappa\kappa}]_{ij} = \begin{cases} E\left[|\kappa(\boldsymbol{u}(\omega_i), \boldsymbol{u}(n))|^2\right], i = j \\ E\left[\kappa(\boldsymbol{u}(\omega_i), \boldsymbol{u}(n)) \cdot \kappa^*\left(\boldsymbol{u}\left(\omega_j\right), \boldsymbol{u}(n)\right)\right], i \neq j \end{cases} \tag{9.24}$$

where $1 \leq i, j \leq M$. The result for the complex kernel is shown here (see Appendix C for the complexified kernel).

Using the complex Gaussian kernel (9.6), the result for (9.24) can be shown to be made up of 2 cases:

**Case (1)** For the $i = j$ case:

$$E\left[|\kappa(u(\omega_i), u(n))|^2\right] = E\left[\exp\left[\frac{-1}{2\xi^2}\left(\sum_{k=1}^{L}\left[u_k(\omega_i) - u_k^*(n)\right]^2 - \sum_{k=1}^{L}\left[u_k(\omega_i) - u_k^*(\omega_i)\right]^2 + \sum_{k=1}^{L}\left[u_k^*(\omega_i) - u_k(n)\right]^2 - \sum_{k=1}^{L}\left[u_k^*(\omega_i) - u_k(\omega_i)\right]^2\right)\right]\right] \tag{9.25}$$

Considering input signal model (9.2), we form vector $\mathbf{y} = [\mathbf{x}_{re}{}^T(n)\, \mathbf{x}_{im}{}^T(n)\, \mathbf{x}_{re}{}^T(\omega_i)\, \mathbf{x}_{im}{}^T(\omega_i)]^T$ and for analyzing (9.25), write

$$\begin{aligned}
&\sum_{k=1}^{L}\left[u_k(\omega_i) - u_k^*(n)\right]^2 - \sum_{k=1}^{L}\left[u_k(\omega_i) - u_k^*(\omega_i)\right]^2 \\
&= \sum_{k=1}^{L}\left[-\rho_{re}x_{re,k}(n) + i\rho_{im}x_{im,k}(n) + \rho_{re}x_{re,k}(\omega_i) + i\rho_{im}x_{im,k}(\omega_i)\right]^2 \\
&\quad - \sum_{k=1}^{L}\left[\rho_{re}x_{re,k}(\omega_i) + i\rho_{im}x_{im,k}(\omega_i) - \rho_{re}x_{re,k}(\omega_i) + i\rho_{im}x_{im,k}(\omega_i)\right]^2 \\
&= \boldsymbol{y}^T\boldsymbol{O}_{A}\boldsymbol{y} - \boldsymbol{y}^T\boldsymbol{O}_{B}\boldsymbol{y}
\end{aligned} \tag{9.26}$$

where

$$\boldsymbol{O}_A = \begin{bmatrix} \rho_{re}^2\boldsymbol{I} & -i\rho_{re}\rho_{im}\boldsymbol{I} & -\rho_{re}^2\boldsymbol{I} & -i\rho_{re}\rho_{im}\boldsymbol{I} \\ -i\rho_{re}\rho_{im}\boldsymbol{I} & -\rho_{im}^2\boldsymbol{I} & i\rho_{re}\rho_{im}\boldsymbol{I} & -\rho_{im}^2\boldsymbol{I} \\ -\rho_{re}^2\boldsymbol{I} & i\rho_{re}\rho_{im}\boldsymbol{I} & \rho_{re}^2\boldsymbol{I} & i\rho_{re}\rho_{im}\boldsymbol{I} \\ -i\rho_{re}\rho_{im}\boldsymbol{I} & -\rho_{im}^2\boldsymbol{I} & i\rho_{re}\rho_{im}\boldsymbol{I} & -\rho_{im}^2\boldsymbol{I} \end{bmatrix}, \boldsymbol{O}_B = \begin{bmatrix} 0\,0\,0 & 0 \\ 0\,0\,0 & 0 \\ 0\,0\,0 & 0 \\ 0\,0\,0 & -4\rho_{im}^2\boldsymbol{I} \end{bmatrix} \tag{9.27}$$

Equation (9.26) can be equivalently expressed as $\mathbf{y}^T\mathbf{O}_{\mathrm{A}}\mathbf{y} - \mathbf{y}^T\mathbf{O}_{\mathrm{B}}\mathbf{y} = \mathbf{y}^T\mathbf{O}_{\mathrm{C}}\mathbf{y}$, where $\mathbf{O}_{\mathrm{C}} = \mathbf{O}_{\mathrm{A}} - \mathbf{O}_{\mathrm{B}}$. Now including the second set of summations from (9.25), we get

$$\boldsymbol{y}^T\boldsymbol{O}_C\boldsymbol{y} + \boldsymbol{y}^T\boldsymbol{O}_C^*\boldsymbol{y} = \boldsymbol{y}^T\boldsymbol{O}_{r1}\boldsymbol{y} \tag{9.28}$$

where $\mathrm{O}_{\mathrm{r}1} = (\mathrm{O}_{\mathrm{C}} + \mathrm{O}_{\mathrm{C}}{}^*)$.

Therefore Eq. (9.25) can be written

$$E\left[|\kappa(\boldsymbol{u}(\omega_i),\boldsymbol{u}(n))|^2\right] = E\left[\exp\left(\frac{-1}{2\xi^2}\boldsymbol{y}^T\boldsymbol{O}_{r1}\boldsymbol{y}\right)\right] \tag{9.29}$$

For evaluating (9.29), we use the following property. Let $z$ be an expression of the form: $z = \mathbf{y}^T\mathbf{O}\mathbf{y}$, where $\mathbf{y}$ is a zero-mean Gaussian vector. The characteristic function of term $z$ may then be expressed [1]:

$$\psi_z(s) = E\left\{e^{sz}\right\} = \det(\boldsymbol{I} - 2s\boldsymbol{O}\boldsymbol{R}_y)^{-1/2} \tag{9.30}$$

where $\mathbf{R_y}$ is the covariance matrix of $\mathbf{y}$. Therefore, using $s = -1/(2\xi^2)$, we can write

$$E\left\{e^{-(y^T O_{r1} y)/(2\xi^2)}\right\} = \det(\boldsymbol{I} + \boldsymbol{O}_{r1}\boldsymbol{R}_y/\xi^2)^{-1/2} \tag{9.31}$$

where

$$\boldsymbol{O}_{r1} = \begin{bmatrix} 2\rho_{re}^2\boldsymbol{I} & 0 & -2\rho_{re}^2\boldsymbol{I} & 0 \\ 0 & -2\rho_{im}^2\boldsymbol{I} & 0 & -2\rho_{im}^2\boldsymbol{I} \\ -2\rho_{re}^2\boldsymbol{I} & 0 & 2\rho_{re}^2\boldsymbol{I} & 0 \\ 0 & -2\rho_{im}^2\boldsymbol{I} & 0 & 6\rho_{im}^2\boldsymbol{I} \end{bmatrix} \tag{9.32}$$

and vector $\mathbf{y}$ described previously. The result for (9.31) can be seen to be real-valued.

**Case (2)** For the $i \neq j$ case:

$$\begin{aligned} &E\left[\kappa(\boldsymbol{u}(\omega_i),\boldsymbol{u}(n))\cdot\kappa^*\left(\boldsymbol{u}(\omega_j),\boldsymbol{u}(n)\right)\right] = \\ &E\left[\exp\left[\frac{-1}{2\xi^2}\left(\sum_{k=1}^{L}\left[u_k(\omega_i) - u_k^*(n)\right]^2 - \sum_{k=1}^{L}\left[u_k(\omega_i) - u_k^*(\omega_i)\right]^2\right.\right.\right. \\ &\left.\left.\left.+\sum_{k=1}^{L}\left[u_k^*(\omega_j) - u_k(n)\right]^2 - \sum_{k=1}^{L}\left[u_k^*(\omega_j) - u_k(\omega_j)\right]^2\right)\right]\right] \end{aligned} \tag{9.33}$$

Forming vector $\mathbf{y} = [\mathbf{x}_{re}{}^T(n)\ \mathbf{x}_{im}{}^T(n)\ \mathbf{x}_{re}{}^T(\omega_i)\ \mathbf{x}_{im}{}^T(\omega_i)\ \mathbf{x}_{re}{}^T(\omega_j)\ \mathbf{x}_{im}{}^T(\omega_j)]^T$, we can write

$$\begin{aligned} &\sum_{k=1}^{L}\left[u_k(\omega_i) - u_k^*(n)\right]^2 - \sum_{k=1}^{L}\left[u_k(\omega_i) - u_k^*(\omega_i)\right]^2 \\ &+\sum_{k=1}^{L}\left[u_k^*(\omega_j) - u_k(n)\right]^2 - \sum_{k=1}^{L}\left[u_k^*(\omega_j) - u_k(\omega_j)\right]^2 = \boldsymbol{y}^T\boldsymbol{O}_{r2}\boldsymbol{y} \end{aligned} \tag{9.34}$$

where $O_{r2} = \begin{bmatrix} 2\rho_{re}^2 I & 0 & -\rho_{re}^2 I & -i\rho_{re}\rho_{im} I & -\rho_{re}^2 I & i\rho_{re}\rho_{im} I \\ 0 & -2\rho_{im}^2 I & i\rho_{re}\rho_{im} I & -\rho_{im}^2 I & -i\rho_{re}\rho_{im} I & -\rho_{im}^2 I \\ -\rho_{re}^2 I & i\rho_{re}\rho_{im} I & \rho_{re}^2 I & i\rho_{re}\rho_{im} I & 0 & 0 \\ -i\rho_{re}\rho_{im} I & -\rho_{im}^2 I & i\rho_{re}\rho_{im} I & 3\rho_{im}^2 I & 0 & 0 \\ -\rho_{re}^2 I & -i\rho_{re}\rho_{im} I & 0 & 0 & \rho_{re}^2 I & -i\rho_{re}\rho_{im} I \\ i\rho_{re}\rho_{im} I & -\rho_{im}^2 I & 0 & 0 & -i\rho_{re}\rho_{im} I & 3\rho_{im}^2 I \end{bmatrix}$

Finally, based on Eq. (9.30) we have

$$E\left\{e^{-(y^T O_{r2} y)/(2\xi^2)}\right\} = \det(I + O_{r2} R_y/\xi^2)^{-1/2} \tag{9.35}$$

Additionally for $i \neq j$, considering the equivalence of the statistics between vectors $\mathbf{x}_{re}(\omega_i)$ and $\mathbf{x}_{re}(\omega_j)$, as well as between $\mathbf{x}_{im}(\omega_i)$ and $\mathbf{x}_{im}(\omega_j)$, we can see that

$$E\left[\kappa(\boldsymbol{u}(\omega_i), \boldsymbol{u}(n)) \cdot \kappa^*\left(\boldsymbol{u}(\omega_j), \boldsymbol{u}(n)\right)\right] = E\left[\kappa\left(\boldsymbol{u}(\omega_j), \boldsymbol{u}(n)\right) \cdot \kappa^*(\boldsymbol{u}(\omega_i), \boldsymbol{u}(n))\right]$$

i.e., the result (9.35) should be real-valued.

We have evaluated each element of matrix $\mathbf{R}_{\kappa\kappa}$ from (9.24) (i.e., the $i = j$ and $i \neq j$ cases).

### 9.2.2 Finding Matrix $C_v(N)$

For finding matrix $\mathbf{C}_v(n)$, writing the weight-error update as (combining (9.21) and (9.22)):

$$\boldsymbol{v}(n+1) = \boldsymbol{v}(n) + \eta e_0^*(n)\boldsymbol{\kappa}_\omega(n) - \eta \boldsymbol{\kappa}_\omega^H(n)\boldsymbol{v}(n)\boldsymbol{\kappa}_\omega(n) \tag{9.36}$$

Then $\mathbf{C}_v(n)$ can be expressed:

$$\begin{aligned} C_v(n+1) &= E\left[\boldsymbol{v}(n+1)\boldsymbol{v}^H(n+1)\right] \\ &= E[\left(\boldsymbol{v}(n) + \eta e_0^*(n)\boldsymbol{\kappa}_\omega(n) - \eta\boldsymbol{\kappa}_\omega(n)\boldsymbol{\kappa}_\omega^H(n)\boldsymbol{v}(n)\right) \\ &\times \left(\boldsymbol{v}^H(n) + \eta e_0(n)\boldsymbol{\kappa}_\omega^H(n) - \eta\boldsymbol{v}^H(n)\boldsymbol{\kappa}_\omega(n)\boldsymbol{\kappa}_\omega^H(n)\right)] \\ &\approx E[\boldsymbol{v}(n)\boldsymbol{v}^H(n) - \eta\boldsymbol{v}(n)\boldsymbol{v}^H(n)\boldsymbol{\kappa}_\omega(n)\boldsymbol{\kappa}_\omega^H(n) - \eta\boldsymbol{\kappa}_\omega(n)\boldsymbol{\kappa}_\omega^H(n)\boldsymbol{v}(n)\boldsymbol{v}^H(n) \\ &+ \eta^2\boldsymbol{\kappa}_\omega(n)\boldsymbol{\kappa}_\omega^H(n)\boldsymbol{v}(n)\boldsymbol{v}^H(n)\boldsymbol{\kappa}_\omega(n)\boldsymbol{\kappa}_\omega^H(n) + \eta^2|e_0(n)|^2\boldsymbol{\kappa}_\omega(n)\boldsymbol{\kappa}_\omega^H(n)] \end{aligned} \tag{9.37}$$

For the last step of (9.37), the **MIA** is used, in addition to the assumptions:

$$E\left[e_0(n)\boldsymbol{\kappa}_\omega(n)\boldsymbol{v}^H(n)\right] \approx E[e_0(n)\boldsymbol{\kappa}_\omega(n)]E\left[\boldsymbol{v}^H(n)\right] = 0 \tag{9.38}$$

$$E\left[e_0(n)\boldsymbol{\kappa}_\omega(n)\boldsymbol{v}^H(n)\boldsymbol{\kappa}_\omega(n)\boldsymbol{\kappa}_\omega^H(n)\right]$$

$$\approx E\left[\boldsymbol{v}^{H}(n)\right]E\left[e_0(n)\boldsymbol{\kappa}_{\omega}(n)\right]\boldsymbol{R}_{\kappa\kappa} = 0 \tag{9.39}$$

based on orthogonality principle $E[e_0(n)\boldsymbol{\kappa}_{\omega}(n)] = 0$.

Using **MIA** again with (9.37), and assuming $\boldsymbol{\kappa}_{\omega}(n)\boldsymbol{\kappa}_{\omega}{}^{H}(n)$ and $e_0(n)$ are uncorrelated (assumed similarly to **MIA**):

$$\boldsymbol{C}_{\boldsymbol{v}}(n+1) \approx \boldsymbol{C}_{\boldsymbol{v}}(n) - \eta\boldsymbol{C}_{\boldsymbol{v}}(n)\boldsymbol{R}_{\kappa\kappa} - \eta\boldsymbol{R}_{\kappa\kappa}\boldsymbol{C}_{\boldsymbol{v}}(n) + \eta^2\boldsymbol{T}(n) + \eta^2 J_{\min}\boldsymbol{R}_{\kappa\kappa} \tag{9.40}$$

where $\mathbf{T}(n)$ is matrix

$$\boldsymbol{T}(n) = E\left[\boldsymbol{\kappa}_{\omega}(n)\boldsymbol{\kappa}_{\omega}^{H}(n)\boldsymbol{v}(n)\boldsymbol{v}^{H}(n)\boldsymbol{\kappa}_{\omega}(n)\boldsymbol{\kappa}_{\omega}^{H}(n)\right] \tag{9.41}$$

with elements

$$[\boldsymbol{T}(n)]_{ij} \approx \sum_{l=1}^{M}\sum_{p=1}^{M} E\left[\kappa_{\omega_i}(n)\kappa_{\omega_l}^{*}(n)\kappa_{\omega_p}(n)\kappa_{\omega_j}^{*}(n)\right][\boldsymbol{C}_{\boldsymbol{v}}(n)]_{lp} \tag{9.42}$$

where $\boldsymbol{\kappa}_{\boldsymbol{\omega}_q}(\boldsymbol{n}) = \kappa(\mathrm{u}(\omega_q), \mathrm{u}(n))$.

The $E[.]$ terms of (9.42) needed for finding $\mathbf{C}_{\mathbf{v}}(n)$ using the complex and complexified kernel are described next.

### 9.2.3 Evaluating $C_v(N)$ with Complexified Kernel

The $E[.]$ terms for various values of $i, j, l, p$ considering the complexified kernel may be found as follows. Note the kernel evaluations for this case are always real-valued from (9.7). Therefore, the conjugation operations in (9.42) may be ignored.

Depending on the values of $i, j, l, p$, each $E[.]$ term of (9.42) may evaluate to one of the following 5 cases.

**Case (1)** For the $i = j = p = l$ case: $E\{\kappa_{\omega i}(n)\kappa_{\omega l}{}^{*}(n)\kappa_{\omega p}(n)\kappa_{\omega j}{}^{*}(n)\} = E\{\kappa_{\omega i}{}^{4}(n)\}$ can be evaluated as

$$E\left[\kappa^4(\boldsymbol{u}(\omega_i), \boldsymbol{u}(n))\right] =$$
$$E\left[\exp\left[\frac{-4}{2\xi^2}\left(\sum_{k=1}^{L}\mathrm{Re}\{u_k(n) - u_k(\omega_i)\}^2 + \sum_{k=1}^{L} Im\{u_k(n) - u_k(\omega_i)\}^2\right)\right]\right] \tag{9.43}$$

If we form vector $\mathbf{y} = [\mathbf{x}_{re}{}^{T}(n)\,\mathbf{x}_{im}{}^{T}(n)\,\mathbf{x}_{re}{}^{T}(\omega_i)\,\mathbf{x}_{im}{}^{T}(\omega_i)]^{T}$, then

$$4\sum_{k=1}^{L}\mathrm{Re}\{u_k(n) - u_k(\omega_i)\}^2 + 4\sum_{k=1}^{L} Im\{u_k(n) - u_k(\omega_i)\}^2 = \boldsymbol{y}^{T}\boldsymbol{O}_1\boldsymbol{y} \tag{9.44}$$

where

$$\boldsymbol{O}_1 = \begin{bmatrix} 4 & -4 \\ -4 & 4 \end{bmatrix} \otimes \begin{bmatrix} \rho_{re}^2 \boldsymbol{I} & 0 \\ 0 & \rho_{im}^2 \boldsymbol{I} \end{bmatrix} \tag{9.45}$$

and $\otimes$ is the Kronecker product. Finally, using (9.30) with $z = \mathbf{y}^T \mathbf{O}_1 \mathbf{y}$ and $s = -1/(2\xi^2)$ yields

$$E\left\{e^{-(y^T O_1 y)/(2\xi^2)}\right\} = \det(\boldsymbol{I} + \boldsymbol{O}_1 \boldsymbol{R}_y/\xi^2)^{-1/2} = \mu_1 \tag{9.46}$$

**Case (2)** For the $i = j = p \neq l$ case: $E\{\kappa_{\omega i}(n)\kappa_{\omega l}{}^*(n)\kappa_{\omega p}(n)\kappa_{\omega j}{}^*(n)\} = E\{\kappa_{\omega i}{}^3(n)\kappa_{\omega l}(n)\}$, which can be evaluated as

$$\begin{aligned} &E\left[\kappa^3(\boldsymbol{u}(\omega_i), \boldsymbol{u}(n)) \cdot \kappa(\boldsymbol{u}(\omega_l), \boldsymbol{u}(n))\right] = \\ &E\left[\exp\left[\frac{-1}{2\xi^2}\left(3\sum_{k=1}^{L} \mathrm{Re}\{u_k(n) - u_k(\omega_i)\}^2 + 3\sum_{k=1}^{L} Im\{u_k(n) - u_k(\omega_i)\}^2\right.\right.\right. \\ &\left.\left.\left. + \sum_{k=1}^{L} \mathrm{Re}\{u_k(n) - u_k(\omega_l)\}^2 + \sum_{k=1}^{L} Im\{u_k(n) - u_k(\omega_l)\}^2\right)\right]\right] \end{aligned} \tag{9.47}$$

Forming vector $\mathbf{y} = [\mathbf{x}_{re}{}^T(n)\ \mathbf{x}_{im}{}^T(n)\ \mathbf{x}_{re}{}^T(\omega_i)\ \mathbf{x}_{im}{}^T(\omega_i)\ \mathbf{x}_{re}{}^T(\omega_l)\ \mathbf{x}_{im}{}^T(\omega_l)]^T$, we can write

$$\begin{aligned} &3\sum_{k=1}^{L} \mathrm{Re}\{u_k(n) - u_k(\omega_i)\}^2 + 3\sum_{k=1}^{L} Im\{u_k(n) - u_k(\omega_i)\}^2 \\ &+ \sum_{k=1}^{L} \mathrm{Re}\{u_k(n) - u_k(\omega_l)\}^2 + \sum_{k=1}^{L} Im\{u_k(n) - u_k(\omega_l)\}^2 = \boldsymbol{y}^T \boldsymbol{O}_2 \boldsymbol{y} \end{aligned} \tag{9.48}$$

where

$$\boldsymbol{O}_2 = \begin{bmatrix} 4 & -3 & -1 \\ -3 & 3 & 0 \\ -1 & 0 & 1 \end{bmatrix} \otimes \begin{bmatrix} \rho_{re}^2 \boldsymbol{I} & 0 \\ 0 & \rho_{im}^2 \boldsymbol{I} \end{bmatrix} \tag{9.49}$$

Which using Eq. (9.30), yields

$$E\left\{e^{-(y^T O_2 y)/(2\xi^2)}\right\} = \det(\boldsymbol{I} + \boldsymbol{O}_2 \boldsymbol{R}_y/\xi^2)^{-1/2} = \mu_2 \tag{9.50}$$

**Case (3)** For the $i = j \neq p = l$ case: $E\{\kappa_{\omega i}(n)\kappa_{\omega l}{}^*(n)\kappa_{\omega p}(n)\kappa_{\omega j}{}^*(n)\} = E\{\kappa_{\omega i}{}^2(n)\kappa_{\omega l}{}^2(n)\}$. Forming $\mathbf{y} = [\mathbf{x}_{re}{}^T(n)\ \mathbf{x}_{im}{}^T(n)\ \mathbf{x}_{re}{}^T(\omega_i)\ \mathbf{x}_{im}{}^T(\omega_i)\ \mathbf{x}_{re}{}^T(\omega_l)$

$\mathbf{x}_{im}{}^T(\omega_l)]^T$, the result can be seen to be

$$E\left\{e^{-(y^T O_3 y)/(2\xi^2)}\right\} = \det(\boldsymbol{I} + \boldsymbol{O}_3\boldsymbol{R}_y/\xi^2)^{-1/2} = \mu_3 \tag{9.51}$$

where

$$\boldsymbol{O}_3 = \begin{bmatrix} 4 & -2 & -2 \\ -2 & 2 & 0 \\ -2 & 0 & 2 \end{bmatrix} \otimes \begin{bmatrix} \rho_{re}^2 \boldsymbol{I} & 0 \\ 0 & \rho_{im}^2 \boldsymbol{I} \end{bmatrix} \tag{9.52}$$

**Case (4)** For the $i = j \neq p \neq l$ case: $E\{\kappa_{\omega i}(n)\kappa_{\omega l}{}^*(n)\kappa_{\omega p}(n)\kappa_{\omega j}{}^*(n)\} = E\{\kappa_{\omega i}{}^2(n)\kappa_{\omega p}(n)\kappa_{\omega l}(n)\}$. Forming $\mathbf{y} = [\mathbf{x}_{re}{}^T(n)\ \mathbf{x}_{im}{}^T(n)\ \mathbf{x}_{re}{}^T(\omega_i)\ \mathbf{x}_{im}{}^T(\omega_i)\ \mathbf{x}_{re}{}^T(\omega_p)\ \mathbf{x}_{im}{}^T(\omega_p)\ \mathbf{x}_{re}{}^T(\omega_l)\ \mathbf{x}_{im}{}^T(\omega_l)]^T$, the result is

$$E\left\{e^{-(y^T O_4 y)/(2\xi^2)}\right\} = \det(\boldsymbol{I} + \boldsymbol{O}_4\boldsymbol{R}_y/\xi^2)^{-1/2} = \mu_4 \tag{9.53}$$

where

$$\boldsymbol{O}_4 = \begin{bmatrix} 4 & -2 & -1 & -1 \\ -2 & 2 & 0 & 0 \\ -1 & 0 & 1 & 0 \\ -1 & 0 & 0 & 1 \end{bmatrix} \otimes \begin{bmatrix} \rho_{re}^2 \boldsymbol{I} & 0 \\ 0 & \rho_{im}^2 \boldsymbol{I} \end{bmatrix} \tag{9.54}$$

**Case (5)** For the $i \neq j \neq p \neq l$ case, we evaluate: $E\{\kappa_{\omega i}(n)\kappa_{\omega l}{}^*(n)\ \kappa_{\omega p}(n)\kappa_{\omega j}{}^*(n)\}$. With $\mathbf{y} = [\mathbf{x}_{re}{}^T(n)\ \mathbf{x}_{im}{}^T(n)\ \mathbf{x}_{re}{}^T(\omega_i)\ \mathbf{x}_{im}{}^T(\omega_i)\ \mathbf{x}_{re}{}^T(\omega_l)\ \mathbf{x}_{im}{}^T(\omega_l)\ \mathbf{x}_{re}{}^T(\omega_p)\ \mathbf{x}_{im}{}^T(\omega_p)\ \mathbf{x}_{re}{}^T(\omega_j)\ \mathbf{x}_{im}{}^T(\omega_j)]^T$, the result is

$$E\left\{e^{-(y^T O_5 y)/(2\xi^2)}\right\} = \det(\boldsymbol{I} + \boldsymbol{O}_5\boldsymbol{R}_y/\xi^2)^{-1/2} = \mu_5 \tag{9.55}$$

where

$$\boldsymbol{O}_5 = \begin{bmatrix} 4 & -1 & -1 & -1 & -1 \\ -1 & 1 & 0 & 0 & 0 \\ -1 & 0 & 1 & 0 & 0 \\ -1 & 0 & 0 & 1 & 0 \\ -1 & 0 & 0 & 0 & 1 \end{bmatrix} \otimes \begin{bmatrix} \rho_{re}^2 \boldsymbol{I} & 0 \\ 0 & \rho_{im}^2 \boldsymbol{I} \end{bmatrix} \tag{9.56}$$

Therefore we evaluated terms $\mu_1$ to $\mu_5$ describing the 5 possible values of any $E[.]$ term of (9.42) using the complexified kernel.

### 9.2.4 Evaluating $C_v(N)$ with Complex Kernel

The $E[.]$ terms of (9.42) considering the complex Gaussian kernel are found as follows. Note there are more cases to consider with the complex kernel than the complexified kernel (denoted with subcases, i.e., $\mu_{2a}$, $\mu_{2b}$, etc.) since the kernel evaluations using the complex Gaussian kernel may be complex-valued. The use of subcases will aid analysis later.

**Case (1)** For the $i = j = p = l$ case: $E\{\kappa_{\omega i}(n)\kappa_{\omega l}{}^*(n)\kappa_{\omega p}(n)\kappa_{\omega j}{}^*(n)\} = E\{|\kappa_{\omega i}(n)|^4\}$ can be evaluated as

$$E\big[|\kappa(\boldsymbol{u}(\omega_i), \boldsymbol{u}(n))|^4\big] =$$
$$E\left[\exp\left[\frac{-2}{2\xi^2}\left(\sum_{k=1}^{L}\big[u_k(\omega_i) - u_k^*(n)\big]^2 - \sum_{k=1}^{L}\big[u_k(\omega_i) - u_k^*(\omega_i)\big]^2 + \sum_{k=1}^{L}[u_k^*(\omega_i) - u_k(n)]^2 - \sum_{k=1}^{L}[u_k^*(\omega_i) - u_k(\omega_i)]^2\right)\right]\right] \tag{9.57}$$

If we form the vector $\mathbf{y} = [\mathbf{x}_{re}{}^T(n)\ \mathbf{x}_{im}{}^T(n)\ \mathbf{x}_{re}{}^T(\omega_i)\ \mathbf{x}_{im}{}^T(\omega_i)]^T$, we can write

$$2\sum_{k=1}^{L}\big[u_k(\omega_i) - u_k^*(n)\big]^2 - 2\sum_{k=1}^{L}\big[u_k(\omega_i) - u_k^*(\omega_i)\big]^2 + 2\sum_{k=1}^{L}\big[u_k^*(\omega_i) - u_k(n)\big]^2 - 2\sum_{k=1}^{L}\big[u_k^*(\omega_i) - u_k(\omega_i)\big]^2 = \mathbf{y}^T\boldsymbol{O}_1\mathbf{y} \tag{9.58}$$

where $\mathbf{O}_1$, considering (9.32), is

$$\boldsymbol{O}_1 = \begin{bmatrix} 4\rho_{re}^2\boldsymbol{I} & 0 & -4\rho_{re}^2\boldsymbol{I} & 0 \\ 0 & -4\rho_{im}^2\boldsymbol{I} & 0 & -4\rho_{im}^2\boldsymbol{I} \\ -4\rho_{re}^2\boldsymbol{I} & 0 & 4\rho_{re}^2\boldsymbol{I} & 0 \\ 0 & -4\rho_{im}^2\boldsymbol{I} & 0 & 12\rho_{im}^2\boldsymbol{I} \end{bmatrix} \tag{9.59}$$

Finally, using (9.30) with $z = \mathbf{y}^T\mathbf{O}_1\mathbf{y}$, we can write

$$E\left\{e^{-(\mathbf{y}^T\boldsymbol{O}_1\mathbf{y})/(2\xi^2)}\right\} = \det(\boldsymbol{I} + \boldsymbol{O}_1\boldsymbol{R}_y/\xi^2)^{-1/2} = \mu_1 \tag{9.60}$$

**Case (2a)** For the $i = j = p \neq l$ case: $E\{\kappa_{\omega i}(n)\kappa_{\omega l}{}^*(n)\kappa_{\omega p}(n)\kappa_{\omega j}{}^*(n)\}$ can be evaluated:

$$E\big[\kappa^2(\boldsymbol{u}(\omega_i), \boldsymbol{u}(n)) \cdot \kappa^*(\boldsymbol{u}(\omega_i), \boldsymbol{u}(n)) \cdot \kappa^*(\boldsymbol{u}(\omega_l), \boldsymbol{u}(n))\big] =$$

$$E\left[\exp\left[\frac{-1}{2\xi^2}\left(2\sum_{k=1}^{L}\left[u_k(\omega_i)-u_k^*(n)\right]^2-2\sum_{k=1}^{L}\left[u_k(\omega_i)-u_k^*(\omega_i)\right]^2\right.\right.\right.$$
$$+\sum_{k=1}^{L}\left[u_k^*(\omega_i)-u_k(n)\right]^2-\sum_{k=1}^{L}\left[u_k^*(\omega_i)-u_k(\omega_i)\right]^2$$
$$\left.\left.\left.+\sum_{k=1}^{L}\left[u_k^*(\omega_l)-u_k(n)\right]^2-\sum_{k=1}^{L}\left[u_k^*(\omega_l)-u_k(\omega_l)\right]^2\right)\right]\right] \tag{9.61}$$

Forming $\mathbf{y}=[\mathbf{x}_{re}{}^T(n)\,\mathbf{x}_{im}{}^T(n)\,\mathbf{x}_{re}{}^T(\omega_i)\,\mathbf{x}_{im}{}^T(\omega_i)\,\mathbf{x}_{re}{}^T(\omega_l)\,\mathbf{x}_{im}{}^T(\omega_l)]^T$, we can show that

$$E\left\{e^{-(\mathbf{y}^T\boldsymbol{O}_{2a}\mathbf{y})/(2\xi^2)}\right\}=\det(\boldsymbol{I}+\boldsymbol{O}_{2a}\boldsymbol{R}_y/\xi^2)^{-1/2}=\mu_{2a} \tag{9.62}$$

where

$$\boldsymbol{O}_{2a}=\begin{bmatrix} 4\rho_{re}^2\boldsymbol{I} & 0 & -3\rho_{re}^2\boldsymbol{I} & -i\rho_{re}\rho_{im}\boldsymbol{I} & -\rho_{re}^2\boldsymbol{I} & i\rho_{re}\rho_{im}\boldsymbol{I} \\ 0 & -4\rho_{im}^2\boldsymbol{I} & i\rho_{re}\rho_{im}\boldsymbol{I} & -3\rho_{im}^2\boldsymbol{I} & -i\rho_{re}\rho_{im}\boldsymbol{I} & -\rho_{im}^2\boldsymbol{I} \\ -3\rho_{re}^2\boldsymbol{I} & i\rho_{re}\rho_{im}\boldsymbol{I} & 3\rho_{re}^2\boldsymbol{I} & i\rho_{re}\rho_{im}\boldsymbol{I} & 0 & 0 \\ -i\rho_{re}\rho_{im}\boldsymbol{I} & -3\rho_{im}^2\boldsymbol{I} & i\rho_{re}\rho_{im}\boldsymbol{I} & 9\rho_{im}^2\boldsymbol{I} & 0 & 0 \\ -\rho_{re}^2\boldsymbol{I} & -i\rho_{re}\rho_{im}\boldsymbol{I} & 0 & 0 & \rho_{re}^2\boldsymbol{I} & -i\rho_{re}\rho_{im}\boldsymbol{I} \\ i\rho_{re}\rho_{im}\boldsymbol{I} & -\rho_{im}^2\boldsymbol{I} & 0 & 0 & -i\rho_{re}\rho_{im}\boldsymbol{I} & 3\rho_{im}^2\boldsymbol{I} \end{bmatrix}$$

**Case (2b)** For the $i=j=l\neq p$ case: $E\{\kappa_{\omega i}(n)\kappa_{\omega l}{}^*(n)\kappa_{\omega p}(n)\kappa_{\omega j}{}^*(n)\}$ can be evaluated:

$$E\left[\kappa^{*2}(\boldsymbol{u}(\omega_i),\boldsymbol{u}(n))\cdot\kappa(\boldsymbol{u}(\omega_i),\boldsymbol{u}(n))\cdot\kappa\left(\boldsymbol{u}\left(\omega_p\right),\boldsymbol{u}(n)\right)\right]$$
$$=E\left[\exp\left[\frac{-1}{2\xi^2}\left(2\sum_{k=1}^{L}[u_k^*(\omega_i)-u_k(n)]^2-2\sum_{k=1}^{L}[u_k^*(\omega_i)-u_k(\omega_i)]^2\right.\right.\right.$$
$$+\sum_{k=1}^{L}[u_k(\omega_i)-u_k^*(n)]^2-\sum_{k=1}^{L}[u_k(\omega_i)-u_k^*(\omega_i)]^2$$
$$\left.\left.\left.+\sum_{k=1}^{L}[u_k\left(\omega_p\right)-u_k^*(n)]^2-\sum_{k=1}^{L}[u_k\left(\omega_p\right)-u_k^*\left(\omega_p\right)]^2\right)\right]\right] \tag{9.63}$$

Forming vector $\mathbf{y}=[\mathbf{x}_{re}{}^T(n)\,\mathbf{x}_{im}{}^T(n)\,\mathbf{x}_{re}{}^T(\omega_i)\,\mathbf{x}_{im}{}^T(\omega_i)\,\mathbf{x}_{re}{}^T(\omega_p)\,\mathbf{x}_{im}{}^T(\omega_p)]^T$, we can show that

$$E\left\{e^{-(\mathbf{y}^T\boldsymbol{O}_{2b}\mathbf{y})/(2\xi^2)}\right\}=\det(\boldsymbol{I}+\boldsymbol{O}_{2b}\boldsymbol{R}_y/\xi^2)^{-1/2}=\mu_{2b} \tag{9.64}$$

where $\mathbf{O}_{2b}=\mathbf{O}_{2a}{}^H$.

**Case (3a)** For the $i = j \neq p = l$ case: $E\{\kappa_{\omega i}(n)\kappa_{\omega l}{}^{*}(n)\kappa_{\omega p}(n)\kappa_{\omega j}{}^{*}(n)\} = E\{|\kappa_{\omega i}(n)|^2|\kappa_{\omega l}(n)|^2\}$. Forming vector $\mathbf{y} = [\mathbf{x}_{re}{}^T(n)\ \mathbf{x}_{im}{}^T(n)\ \mathbf{x}_{re}{}^T(\omega_i)\ \mathbf{x}_{im}{}^T(\omega_i)\ \mathbf{x}_{re}{}^T(\omega_l)\ \mathbf{x}_{im}{}^T(\omega_l)]^T$, the result is

$$E\left\{e^{-(y^T \boldsymbol{O}_{3a}\mathbf{y})/(2\xi^2)}\right\} = \det(\boldsymbol{I} + \boldsymbol{O}_{3a}\boldsymbol{R}_y/\xi^2)^{-1/2} = \mu_{3a} \tag{9.65}$$

where

$$\boldsymbol{O}_{3a} = \begin{bmatrix} 4\rho_{re}^2\boldsymbol{I} & 0 & -2\rho_{re}^2\boldsymbol{I} & 0 & -2\rho_{re}^2\boldsymbol{I} & 0 \\ 0 & -4\rho_{im}^2\boldsymbol{I} & 0 & -2\rho_{im}^2\boldsymbol{I} & 0 & -2\rho_{im}^2\boldsymbol{I} \\ -2\rho_{re}^2\boldsymbol{I} & 0 & 2\rho_{re}^2\boldsymbol{I} & 0 & 0 & 0 \\ 0 & -2\rho_{im}^2\boldsymbol{I} & 0 & 6\rho_{im}^2\boldsymbol{I} & 0 & 0 \\ -2\rho_{re}^2\boldsymbol{I} & 0 & 0 & 0 & 2\rho_{re}^2\boldsymbol{I} & 0 \\ 0 & -2\rho_{im}^2\boldsymbol{I} & 0 & 0 & 0 & 6\rho_{im}^2\boldsymbol{I} \end{bmatrix}$$

**Case (3b)** For the $i = p \neq j = l$ case: $E\{\kappa_{\omega i}(n)\kappa_{\omega l}{}^{*}(n)\kappa_{\omega p}(n)\kappa_{\omega j}{}^{*}(n)\} = E\{\kappa_{\omega i}{}^2(n)\kappa_{\omega l}{}^{*2}(n)\}$. Forming the vector $\mathbf{y} = [\mathbf{x}_{re}{}^T(n)\ \mathbf{x}_{im}{}^T(n)\ \mathbf{x}_{re}{}^T(\omega_i)\ \mathbf{x}_{im}{}^T(\omega_i)\ \mathbf{x}_{re}{}^T(\omega_l)\ \mathbf{x}_{im}{}^T(\omega_l)]^T$, the result is

$$E\left\{e^{-(y^T \boldsymbol{O}_{3b}\mathbf{y})/(2\xi^2)}\right\} = \det(\boldsymbol{I} + \boldsymbol{O}_{3b}\boldsymbol{R}_y/\xi^2)^{-1/2} = \mu_{3b} \tag{9.66}$$

where

$$\boldsymbol{O}_{3b} = \begin{bmatrix} 4\rho_{re}^2\boldsymbol{I} & 0 & -2\rho_{re}^2\boldsymbol{I} & -2i\rho_{re}\rho_{im}\boldsymbol{I} & -2\rho_{re}^2\boldsymbol{I} & 2i\rho_{re}\rho_{im}\boldsymbol{I} \\ 0 & -4\rho_{im}^2\boldsymbol{I} & 2i\rho_{re}\rho_{im}\boldsymbol{I} & -2\rho_{im}^2\boldsymbol{I} & -2i\rho_{re}\rho_{im}\boldsymbol{I} & -2\rho_{im}^2\boldsymbol{I} \\ -2\rho_{re}^2\boldsymbol{I} & 2i\rho_{re}\rho_{im}\boldsymbol{I} & 2\rho_{re}^2\boldsymbol{I} & 2i\rho_{re}\rho_{im}\boldsymbol{I} & 0 & 0 \\ -2i\rho_{re}\rho_{im}\boldsymbol{I} & -2\rho_{im}^2\boldsymbol{I} & 2i\rho_{re}\rho_{im}\boldsymbol{I} & 6\rho_{im}^2\boldsymbol{I} & 0 & 0 \\ -2\rho_{re}^2\boldsymbol{I} & -2i\rho_{re}\rho_{im}\boldsymbol{I} & 0 & 0 & 2\rho_{re}^2\boldsymbol{I} & -2i\rho_{re}\rho_{im}\boldsymbol{I} \\ 2i\rho_{re}\rho_{im}\boldsymbol{I} & -2\rho_{im}^2\boldsymbol{I} & 0 & 0 & -2i\rho_{re}\rho_{im}\boldsymbol{I} & 6\rho_{im}^2\boldsymbol{I} \end{bmatrix}$$

**Case (4a)** For $i = j \neq p \neq l$ case: $E\{\kappa_{\omega i}(n)\kappa_{\omega l}{}^{*}(n)\kappa_{\omega p}(n)\kappa_{\omega j}{}^{*}(n)\} = E\{|\kappa_{\omega i}(n)|^2\kappa_{\omega p}(n)\kappa_{\omega l}{}^{*}(n)\}$. Forming $\mathbf{y} = [\mathbf{x}_{re}{}^T(n)\ \mathbf{x}_{im}{}^T(n)\ \mathbf{x}_{re}{}^T(\omega_i)\ \mathbf{x}_{im}{}^T(\omega_i)\ \mathbf{x}_{re}{}^T(\omega_p)\ \mathbf{x}_{im}{}^T(\omega_p)\ \mathbf{x}_{re}{}^T(\omega_l)\ \mathbf{x}_{im}{}^T(\omega_l)]^T$, the result is

$$E\left\{e^{-(y^T \boldsymbol{O}_{4a}\mathbf{y})/(2\xi^2)}\right\} = \det(\boldsymbol{I} + \boldsymbol{O}_{4a}\boldsymbol{R}_y/\xi^2)^{-1/2} = \mu_{4a} \tag{9.67}$$

where $O_{4a} = \begin{bmatrix} 4\rho_{re}^2\boldsymbol{I} & 0 & -2\rho_{re}^2\boldsymbol{I} & 0 & -\rho_{re}^2\boldsymbol{I} & -i\rho_{re}\rho_{im}\boldsymbol{I} & -\rho_{re}^2\boldsymbol{I} & i\rho_{re}\rho_{im}\boldsymbol{I} \\ 0 & -4\rho_{im}^2\boldsymbol{I} & 0 & -2\rho_{im}^2\boldsymbol{I} & i\rho_{re}\rho_{im}\boldsymbol{I} & -\rho_{im}^2\boldsymbol{I} & -i\rho_{re}\rho_{im}\boldsymbol{I} & -\rho_{im}^2\boldsymbol{I} \\ -2\rho_{re}^2\boldsymbol{I} & 0 & 2\rho_{re}^2\boldsymbol{I} & 0 & 0 & 0 & 0 & 0 \\ 0 & -2\rho_{im}^2\boldsymbol{I} & 0 & 6\rho_{im}^2\boldsymbol{I} & 0 & 0 & 0 & 0 \\ -\rho_{re}^2\boldsymbol{I} & i\rho_{re}\rho_{im}\boldsymbol{I} & 0 & 0 & \rho_{re}^2\boldsymbol{I} & i\rho_{re}\rho_{im}\boldsymbol{I} & 0 & 0 \\ -i\rho_{re}\rho_{im}\boldsymbol{I} & -\rho_{im}^2\boldsymbol{I} & 0 & 0 & i\rho_{re}\rho_{im}\boldsymbol{I} & 3\rho_{im}^2\boldsymbol{I} & 0 & 0 \\ -\rho_{re}^2\boldsymbol{I} & -i\rho_{re}\rho_{im}\boldsymbol{I} & 0 & 0 & 0 & 0 & \rho_{re}^2\boldsymbol{I} & -i\rho_{re}\rho_{im}\boldsymbol{I} \\ i\rho_{re}\rho_{im}\boldsymbol{I} & -\rho_{im}^2\boldsymbol{I} & 0 & 0 & 0 & 0 & -i\rho_{re}\rho_{im}\boldsymbol{I} & 3\rho_{im}^2\boldsymbol{I} \end{bmatrix}$

**Case (4b)** For $i = p \neq j \neq l$ case: $E\{\kappa_{\omega i}(n)\kappa_{\omega l}{}^*(n)\kappa_{\omega p}(n)\kappa_{\omega j}{}^*(n)\} = E\{\kappa_{\omega i}{}^2(n)\kappa_{\omega j}{}^*(n)\kappa_{\omega l}{}^*(n)\}$. Forming $\mathbf{y} = [\mathbf{x}_{re}{}^T(n)\ \mathbf{x}_{im}{}^T(n)\ \mathbf{x}_{re}{}^T(\omega_i)\ \mathbf{x}_{im}{}^T(\omega_i)\ \mathbf{x}_{re}{}^T(\omega_j)\ \mathbf{x}_{im}{}^T(\omega_j)\ \mathbf{x}_{re}{}^T(\omega_l)\ \mathbf{x}_{im}{}^T(\omega_l)]^T$, the result is

$$E\left\{e^{-(\boldsymbol{y}^T\boldsymbol{O}_{4b}\boldsymbol{y})/(2\xi^2)}\right\} = \det(\boldsymbol{I} + \boldsymbol{O}_{4b}\boldsymbol{R}_{\mathbf{y}}/\xi^2)^{-1/2} = \mu_{4b} \tag{9.68}$$

where $\boldsymbol{O}_{4b} = \begin{bmatrix} 4\rho_{re}^2\boldsymbol{I} & 0 & -2\rho_{re}^2\boldsymbol{I} & -2i\rho_{re}\rho_{im}\boldsymbol{I} & -\rho_{re}^2\boldsymbol{I} & i\rho_{re}\rho_{im}\boldsymbol{I} & -\rho_{re}^2\boldsymbol{I} & i\rho_{re}\rho_{im}\boldsymbol{I} \\ 0 & -4\rho_{im}^2\boldsymbol{I} & 2i\rho_{re}\rho_{im}\boldsymbol{I} & -2\rho_{im}^2\boldsymbol{I} & -i\rho_{re}\rho_{im}\boldsymbol{I} & -\rho_{im}^2\boldsymbol{I} & -i\rho_{re}\rho_{im}\boldsymbol{I} & -\rho_{im}^2\boldsymbol{I} \\ -2\rho_{re}^2\boldsymbol{I} & 2i\rho_{re}\rho_{im}\boldsymbol{I} & 2\rho_{re}^2\boldsymbol{I} & 2i\rho_{re}\rho_{im}\boldsymbol{I} & 0 & 0 & 0 & 0 \\ -2i\rho_{re}\rho_{im}\boldsymbol{I} & -2\rho_{im}^2\boldsymbol{I} & 2i\rho_{re}\rho_{im}\boldsymbol{I} & 6\rho_{im}^2\boldsymbol{I} & 0 & 0 & 0 & 0 \\ -\rho_{re}^2\boldsymbol{I} & -i\rho_{re}\rho_{im}\boldsymbol{I} & 0 & 0 & \rho_{re}^2\boldsymbol{I} & -i\rho_{re}\rho_{im}\boldsymbol{I} & 0 & 0 \\ i\rho_{re}\rho_{im}\boldsymbol{I} & -\rho_{im}^2\boldsymbol{I} & 0 & 0 & -i\rho_{re}\rho_{im}\boldsymbol{I} & 3\rho_{im}^2\boldsymbol{I} & 0 & 0 \\ -\rho_{re}^2\boldsymbol{I} & -i\rho_{re}\rho_{im}\boldsymbol{I} & 0 & 0 & 0 & 0 & \rho_{re}^2\boldsymbol{I} & -i\rho_{re}\rho_{im}\boldsymbol{I} \\ i\rho_{re}\rho_{im}\boldsymbol{I} & -\rho_{im}^2\boldsymbol{I} & 0 & 0 & 0 & 0 & -i\rho_{re}\rho_{im}\boldsymbol{I} & 3\rho_{im}^2\boldsymbol{I} \end{bmatrix}$

**Case (4c)** For $j = l \neq i \neq p$ case: $E\{\kappa_{\omega i}(n)\kappa_{\omega l}{}^*(n)\kappa_{\omega p}(n)\kappa_{\omega j}{}^*(n)\} = E\{\kappa_{\omega j}{}^{*2}(n)\kappa_{\omega i}(n)\kappa_{\omega p}(n)\}$. Forming $\mathbf{y} = [\mathbf{x}_{re}{}^T(n)\ \mathbf{x}_{im}{}^T(n)\ \mathbf{x}_{re}{}^T(\omega_j)\ \mathbf{x}_{im}{}^T(\omega_j)\ \mathbf{x}_{re}{}^T(\omega_i)\ \mathbf{x}_{im}{}^T(\omega_i)\ \mathbf{x}_{re}{}^T(\omega_p)\ \mathbf{x}_{im}{}^T(\omega_p)]^T$, the result is

$$E\left\{e^{-(\boldsymbol{y}^T\boldsymbol{O}_{4c}\boldsymbol{y})/(2\xi^2)}\right\} = \det(\boldsymbol{I} + \boldsymbol{O}_{4c}\boldsymbol{R}_{\mathbf{y}}/\xi^2)^{-1/2} = \mu_{4c} \tag{9.69}$$

where $\mathbf{O}_{4c} = \mathbf{O}_{4b}{}^H$.

**Case (5)** For the $i \neq j \neq p \neq l$ case, we evaluate $E\{\kappa_{\omega i}(n)\kappa_{\omega l}{}^*(n)\ \kappa_{\omega p}(n)\kappa_{\omega j}{}^*(n)\}$. With $\mathbf{y} = [\mathbf{x}_{re}{}^T(n)\ \mathbf{x}_{im}{}^T(n)\ \mathbf{x}_{re}{}^T(\omega_i)\ \mathbf{x}_{im}{}^T(\omega_i)\ \mathbf{x}_{re}{}^T(\omega_l)\ \mathbf{x}_{im}{}^T(\omega_l)\ \mathbf{x}_{re}{}^T(\omega_p)\ \mathbf{x}_{im}{}^T(\omega_p)\ \mathbf{x}_{re}{}^T(\omega_j)\ \mathbf{x}_{im}{}^T(\omega_j)]^T$, the result is

$$E\left\{e^{-(\boldsymbol{y}^T\boldsymbol{O}_5\boldsymbol{y})/(2\xi^2)}\right\} = \det(\boldsymbol{I} + \boldsymbol{O}_5\boldsymbol{R}_{\mathbf{y}}/\xi^2)^{-1/2} = \mu_5 \tag{9.70}$$

where

$$o_5 = \begin{bmatrix}
4\rho_{re}^2\boldsymbol{I} & 0 & -\rho_{re}^2\boldsymbol{I} & -i\rho_{re}\rho_{im}\boldsymbol{I} & -\rho_{re}^2\boldsymbol{I} & i\rho_{re}\rho_{im}\boldsymbol{I} & -\rho_{re}^2\boldsymbol{I} & -i\rho_{re}\rho_{im}\boldsymbol{I} & -\rho_{re}^2\boldsymbol{I} & i\rho_{re}\rho_{im}\boldsymbol{I} \\
0 & -4\rho_{im}^2\boldsymbol{I} & i\rho_{re}\rho_{im}\boldsymbol{I} & -\rho_{im}^2\boldsymbol{I} & -i\rho_{re}\rho_{im}\boldsymbol{I} & -\rho_{im}^2\boldsymbol{I} & i\rho_{re}\rho_{im}\boldsymbol{I} & -\rho_{im}^2\boldsymbol{I} & -i\rho_{re}\rho_{im}\boldsymbol{I} & -\rho_{im}^2\boldsymbol{I} \\
-\rho_{re}^2\boldsymbol{I} & i\rho_{re}\rho_{im}\boldsymbol{I} & \rho_{re}^2\boldsymbol{I} & i\rho_{re}\rho_{im}\boldsymbol{I} & 0 & 0 & 0 & 0 & 0 & 0 \\
-i\rho_{re}\rho_{im}\boldsymbol{I} & -\rho_{im}^2\boldsymbol{I} & i\rho_{re}\rho_{im}\boldsymbol{I} & 3\rho_{im}^2\boldsymbol{I} & 0 & 0 & 0 & 0 & 0 & 0 \\
-\rho_{re}^2\boldsymbol{I} & -i\rho_{re}\rho_{im}\boldsymbol{I} & 0 & 0 & \rho_{re}^2\boldsymbol{I} & -i\rho_{re}\rho_{im}\boldsymbol{I} & 0 & 0 & 0 & 0 \\
i\rho_{re}\rho_{im}\boldsymbol{I} & -\rho_{im}^2\boldsymbol{I} & 0 & 0 & -i\rho_{re}\rho_{im}\boldsymbol{I} & 3\rho_{im}^2\boldsymbol{I} & 0 & 0 & 0 & 0 \\
-\rho_{re}^2\boldsymbol{I} & i\rho_{re}\rho_{im}\boldsymbol{I} & 0 & 0 & 0 & 0 & \rho_{re}^2\boldsymbol{I} & i\rho_{re}\rho_{im}\boldsymbol{I} & 0 & 0 \\
-i\rho_{re}\rho_{im}\boldsymbol{I} & -\rho_{im}^2\boldsymbol{I} & 0 & 0 & 0 & 0 & i\rho_{re}\rho_{im}\boldsymbol{I} & 3\rho_{im}^2\big|_{I}\boldsymbol{I} & 0 & 0 \\
-\rho_{re}^2\boldsymbol{I} & -i\rho_{re}\rho_{im}\boldsymbol{I} & 0 & 0 & 0 & 0 & 0 & 0 & \rho_{re}^2\boldsymbol{I} & -i\rho_{re}\rho_{im}\boldsymbol{I} \\
i\rho_{re}\rho_{im}\boldsymbol{I} & -\rho_{im}^2\boldsymbol{I} & 0 & 0 & 0 & 0 & 0 & 0 & -i\rho_{re}\rho_{im}\boldsymbol{I} & 3\rho_{im}^2\boldsymbol{I}
\end{bmatrix}$$

### 9.2.5 Expressions for $C_v(N)$ and MSE

Using $\mu_1$ to $\mu_5$ (either for the complex Gaussian kernel, or complexified Gaussian), and considering Eq. (9.42), the elements of $\mathbf{T}(n)$ can be expressed as

$$[\boldsymbol{T}(n)]_{ii} = \mu_1[\boldsymbol{C}_v(n)]_{ii} + \sum_{\substack{l=1 \\ l \neq i}}^{M} \left\{ \mu_{2a}[\boldsymbol{C}_v(n)]_{li} + \mu_{2b}[\boldsymbol{C}_v(n)]_{il} + \mu_{3a}[\boldsymbol{C}_v(n)]_{ll} + \mu_{4a} \sum_{\substack{p=1 \\ p \neq \{i,l\}}}^{M} [\boldsymbol{C}_v(n)]_{lp} \right\} \tag{9.71}$$

when $i = j$, and otherwise

$$\begin{aligned}
[\boldsymbol{T}(n)]_{ij} =& \mu_{2a}[\boldsymbol{C}_v(n)]_{ii} + \mu_{2b}[\boldsymbol{C}_v(n)]_{jj} + \mu_{3a}[\boldsymbol{C}_v(n)]_{ij} + \mu_{3b}[\boldsymbol{C}_v(n)]_{ji} \\
&+ \sum_{\substack{l=1 \\ l \neq \{i,j\}}}^{M} \{\mu_{4a}\big([\boldsymbol{C}_v(n)]_{il} + [\boldsymbol{C}_v(n)]_{lj} + [\boldsymbol{C}_v(n)]_{ll}\big) \\
&+ \mu_{4b}[\boldsymbol{C}_v(n)]_{li} + \mu_{4c}[\boldsymbol{C}_v(n)]_{jl} + \mu_5 \sum_{\substack{p=1 \\ p \neq \{i,j,l\}}}^{M} [\boldsymbol{C}_v(n)]_{lp}\}
\end{aligned} \tag{9.72}$$

For the complexified case $\mu_2$, $\mu_3$, and $\mu_4$ are substituted for terms of the form $\mu_{2x}$, $\mu_{3x}$, and $\mu_{4x}$, respectively.

Finally, substituting (9.71), (9.72) into weight-error covariance matrix $\mathbf{C}_v(n)$ of (9.40) yields

$$
\begin{aligned}
[\boldsymbol{C}_{\boldsymbol{v}}(n+1)]_{ii} =& \left(1 - 2\eta r_{md} + \eta^2\mu_1\right)[\boldsymbol{C}_{\boldsymbol{v}}(n)]_{ii} + \eta^2\mu_{3a} \sum_{\substack{l=1\\ l\neq i}}^{M} [\boldsymbol{C}_{\boldsymbol{v}}(n)]_{ll} \\
&+ \left(\eta^2\mu_{2a} - \eta r_{od}\right) \sum_{\substack{l=1\\ l\neq i}}^{M} [\boldsymbol{C}_{\boldsymbol{v}}(n)]_{li} + \left(\eta^2\mu_{2b} - \eta r_{od}\right) \sum_{\substack{l=1\\ l\neq i}}^{M} [\boldsymbol{C}_{\boldsymbol{v}}(n)]_{il} \\
&+ \eta^2\mu_{4a} \sum_{\substack{l=1\\ l\neq i}}^{M} \sum_{\substack{p=1\\ p\neq \{i,l\}}}^{M} [\boldsymbol{C}_{\boldsymbol{v}}(n)]_{lp} + \eta^2 r_{md} J_{\min}
\end{aligned}
\tag{9.73}
$$

when $i = j$, and otherwise

$$
\begin{aligned}
[\boldsymbol{C}_{\boldsymbol{v}}(n+1)]_{ij} =& \left(1 - 2\eta r_{md} + \eta^2\mu_{3a}\right)[\boldsymbol{C}_{\boldsymbol{v}}(n)]_{ij} + \eta^2\mu_{3b}[\boldsymbol{C}_{\boldsymbol{v}}(n)]_{ji} \\
&+ \left(\eta^2\mu_{2a} - \eta r_{od}\right)[\boldsymbol{C}_{\boldsymbol{v}}(n)]_{ii} + \left(\eta^2\mu_{2b} - \eta r_{od}\right)[\boldsymbol{C}_{\boldsymbol{v}}(n)]_{jj} \\
&+ \left(\eta^2\mu_{4a} - \eta r_{od}\right) \sum_{\substack{l=1\\ l\neq \{i,j\}}}^{M} \left([\boldsymbol{C}_{\boldsymbol{v}}(n)]_{il} + [\boldsymbol{C}_{\boldsymbol{v}}(n)]_{lj}\right) \\
&+ \eta^2\mu_{4a} \sum_{\substack{l=1\\ l\neq \{i,j\}}}^{M} [\boldsymbol{C}_{\boldsymbol{v}}(n)]_{ll} + \eta^2\mu_{4b} \sum_{\substack{l=1\\ l\neq \{i,j\}}}^{M} [\boldsymbol{C}_{\boldsymbol{v}}(n)]_{li} \\
&+ \eta^2\mu_{4c} \sum_{\substack{l=1\\ l\neq \{i,j\}}}^{M} [\boldsymbol{C}_{\boldsymbol{v}}(n)]_{jl} \\
&+ \eta^2\mu_5 \sum_{\substack{l=1\\ l\neq \{i,j\}}}^{M} \sum_{\substack{p=1\\ p\neq \{i,j,l\}}}^{M} [\boldsymbol{C}_{\boldsymbol{v}}(n)]_{lp} + \eta^2 r_{od} J_{\min}
\end{aligned}
\tag{9.74}
$$

where $r_{\text{md}} = [\mathbf{R}_{\kappa\kappa}]_{ii}$, and $r_{\text{od}} = [\mathbf{R}_{\kappa\kappa}]_{ij}$ for $i \neq j$. Matrix $\mathbf{R}_{\kappa\kappa}$ is described using either (9.31), (9.35) for the complex Gaussian kernel or (D.6), (D.10) for the complexified form.

For the $\mathbf{C}_{\mathbf{v}}(n)$ expressions, if the complexified form is considered, (9.73) and (9.74) become

$$[\boldsymbol{C}_{\boldsymbol{v}}(n+1)]_{ii} =$$

$$\left(1 - 2\eta r_{md} + \eta^2\mu_1\right)[\boldsymbol{C}_{\boldsymbol{v}}(n)]_{ii} + \eta^2\mu_3 \sum_{\substack{l=1 \\ l \neq i}}^{M} [\boldsymbol{C}_{\boldsymbol{v}}(n)]_{ll}$$

$$+ \left(2\eta^2\mu_2 - 2\eta r_{od}\right) \sum_{\substack{l=1 \\ l \neq i}}^{M} [\boldsymbol{C}_{\boldsymbol{v}}(n)]_{il} + \eta^2\mu_4 \sum_{\substack{l=1 \\ l \neq i}}^{M} \sum_{\substack{p=1 \\ p \neq \{i,l\}}}^{M} [\boldsymbol{C}_{\boldsymbol{v}}(n)]_{lp} + \eta^2 r_{md} J_{\min} \tag{9.75}$$

when $i = j$, and otherwise

$$\begin{aligned}[\boldsymbol{C}_{\boldsymbol{v}}(n+1)]_{ij} =& \left(1 - 2\eta r_{md} + 2\eta^2\mu_3\right)[\boldsymbol{C}_{\boldsymbol{v}}(n)]_{ij} + \eta^2\mu_4 \sum_{\substack{l=1 \\ l \neq \{i,j\}}}^{M} [\boldsymbol{C}_{\boldsymbol{v}}(n)]_{ll} \\ &+ \left(\eta^2\mu_2 - \eta r_{od}\right)\left([\boldsymbol{C}_{\boldsymbol{v}}(n)]_{ii} + [\boldsymbol{C}_{\boldsymbol{v}}(n)]_{jj}\right) \\ &+ \left(2\eta^2\mu_4 - \eta r_{od}\right) \sum_{\substack{l=1 \\ l \neq \{i,j\}}}^{M} \left([\boldsymbol{C}_{\boldsymbol{v}}(n)]_{il} + [\boldsymbol{C}_{\boldsymbol{v}}(n)]_{jl}\right) \\ &+ \eta^2\mu_5 \sum_{\substack{l=1 \\ l \neq \{i,j\}}}^{M} \sum_{\substack{p=1 \\ p \neq \{i,j,l\}}}^{M} [\boldsymbol{C}_{\boldsymbol{v}}(n)]_{lp} + \eta^2 r_{od} J_{\min} \end{aligned} \tag{9.76}$$

which reduces to the expressions of [1] for KLMS (aside from $\mu_1$ to $\mu_5$ differing based on the complexified kernel). Therefore the same stability and steady-state analysis obtained there may be considered for the complexified case (summarized in Appendix D). Combining the $\mathbf{C_v}(n)$, $\mathbf{R_{\kappa\kappa}}$ expressions with Eq. (9.23) yields the MSE equations.

## 9.3 Comparison of Complex and Complexified Kernels

The difference for Eqs. (9.73), (9.74) with the complex kernel is the added subcases (i.e., the $\mu_{2x}$, $\mu_{3x}$, $\mu_{4x}$ cases) compared with $\mu_2$, $\mu_3$, $\mu_4$ of the complexified form. An extensive analysis of the difference between the MSE equations for the complex and complexified kernels remains for future work. However, considering terms $\mu_1$ to $\mu_5$, as well as input circularity $\rho$, the following may be seen:

- When $\rho = 0$, the input is real-valued (i.e., $\rho_{re} = 1$, $\rho_{im} = 0$) and the complex, complexified kernels may be seen to perform the same (i.e., $\mu_1$, $\mu_{2x}$, $\mu_{3x}$, $\mu_{4x}$, $\mu_5$ for the complex kernel become same as $\mu_1$ to $\mu_5$ of complexified). Both also reduce to the results of the KLMS [1] using real inputs/outputs.
- When $\rho = 1$, the input is imaginary (i.e., $\rho_{re} = 0$, $\rho_{im} = 1$). Here the complex and complexified kernels differ. For the complexified, the MSE performance using purely imaginary input/output is symmetric with using real input/output. The complex kernel can be seen to perform differently, however (based on $\mu_1$ to $\mu_5$).

For values of $\rho$ in between (i.e., increasing circularity), the performance based on complex and complexified kernels may be seen to differ also. Terms $\mu_1$ to $\mu_5$ may be complex-valued with the complex kernel form, while they are always real-valued with the complexified. The analysis is seen to be consistent with simulations in Sect. 9.4.

## 9.4 Simulation Results

Here we verify the analysis of the previous section based on simulations. The test method used follows that of [1]. Each simulation test uses 500 ensemble-averaged realizations. For each realization, a dictionary was constructed using the initial 500 samples of sequence $\mathbf{u}(n)$, based on coherence-based sparsification rule:

$$\max_j \left| \kappa(\boldsymbol{u}(l), \boldsymbol{u}(\omega_j)) \right| \le \varepsilon_0 \tag{9.77}$$

An input sample $\mathbf{u}(l)$ was added to the dictionary if (9.77) is met, where parameter $\varepsilon_0$ specifies the dictionary coherence. After the initial 500 samples, the dictionary size $M$ remains constant, and the CKLMS is run for 1000 iterations. During these iterations, the dictionary was updated such that the least recent dictionary element is replaced with input $\mathbf{u}(n\text{-}1)$. The average $M$ across all realizations is used for the theoretical model.

For simulating the system, the input is a sequence of i.i.d. Gaussian vectors following (9.1), i.e.,

$$\boldsymbol{u}(n) = \sqrt{1 - \rho^2} \cdot \boldsymbol{x}_{re}(n) + i\rho \boldsymbol{x}_{im}(n) \tag{9.78}$$

where various values of $\rho$ were used for testing.

The vectors $\mathbf{x}_{re}(n)$, $\mathbf{x}_{im}(n)$ for (9.78) are specified as follows. For both vectors, the vector components are defined $\mathbf{x}(n) = [x_1(n)\ x_2(n)]^T$, and samples $x_1(n)$ and $x_2(n)$ chosen such that $x_1(n) = 0.5x_2(n) + n_x(n)$, where $x_2(n)$ and $n_x(n)$ are independent Gaussian sequences chosen so that ${\sigma_{x1}}^2 = {\sigma_{x2}}^2 = 0.0156$. The simulation test settings described above were used to generate the following examples:

**Example 1** For this example (used in Figs. 9.1, 9.2, 9.3, 9.4, 9.5, 9.6, 9.7 and 9.8, the nonlinear channel followed [1], and consisted of the linear portion.

$$y(n) = [10.5] \cdot \boldsymbol{u}(n) - 0.2y(n-1) + 0.35y(n-2) \tag{9.79}$$

and the nonlinear portion $\psi(\cdot)$ is

$$\psi(y(n)) = \begin{cases} \dfrac{y(n)}{3[0.1 + 0.9y^2(n)]^{1/2}}, \ for\ y(n) \geq 0 \\ \dfrac{-y^2(n)[1-\exp(0.7y(n))]}{3}, \ for\ y(n) < 0 \end{cases} \tag{9.80}$$

The desired response was $d(n) = \psi(y(n)) + z(n)$, where $z(n)$ is zero-mean Gaussian with variance $\sigma_z{}^2 = 10^{-6}$.

For the CKLMS testing, parameters used were $\varepsilon_0 = 5 \times 10^{-1}$ (coherence parameter), and kernel parameter $\xi \in \{0.25, 0.30, 0.35, 0.40\}$. For each $\xi$, dictionary size $M$ was found based on the average across all 500 realizations, as described previously. The following tests were performed.

CKLMS with Complex Gaussian Kernel:
Fig. 9.1: MSE curves: $\eta = 0.072$, $\xi = 0.25$, various $\rho$.
Fig. 9.2: MSE curves: $\eta = 0.072$, $\xi = 0.30$, various $\rho$.
Fig. 9.3: MSE curves: $\eta = 0.072$, $\xi = 0.35$, various $\rho$.
Fig. 9.4: MSE curves: $\eta = 0.072$, $\xi = 0.40$, various $\rho$.
CKLMS with Complexified Gaussian Kernel:
Fig. 9.5: MSE curves: $\eta = 0.072$, $\xi = 0.25$, various $\rho$.
Fig. 9.6: MSE curves: $\eta = 0.072$, $\xi = 0.30$, various $\rho$.
Fig. 9.7: MSE curves: $\eta = 0.072$, $\xi = 0.35$, various $\rho$.
Fig. 9.8: MSE curves: $\eta = 0.072$, $\xi = 0.40$, various $\rho$.

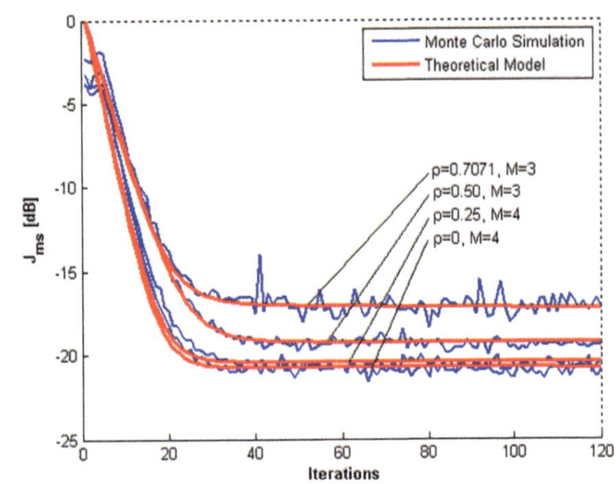


**Fig. 9.1** CKLMS, complex kernel ($\xi = 0.25$, various $\rho$)

Fig. 9.2 CKLMS, complex kernel ($\xi = 0.30$, various $\rho$)

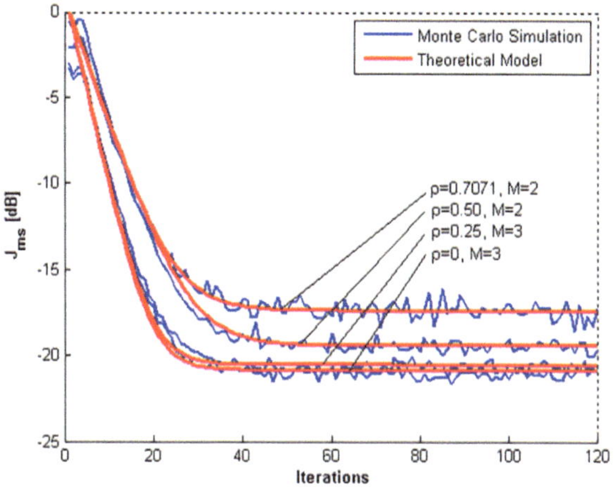


Fig. 9.3 CKLMS, complex kernel ($\xi = 0.35$, various $\rho$)

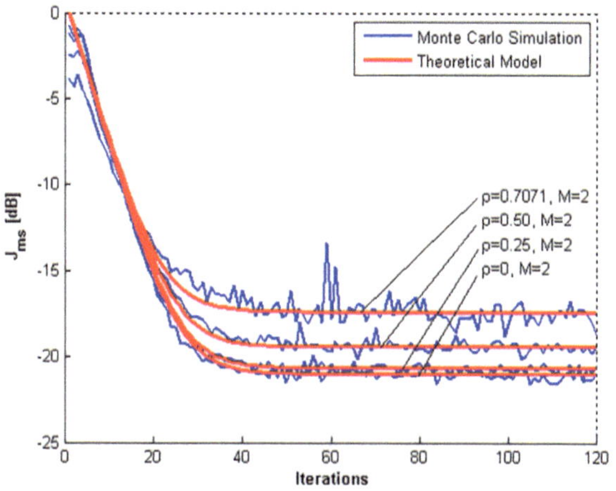


Fig. 9.4 CKLMS, complex kernel ($\xi = 0.40$, various $\rho$)

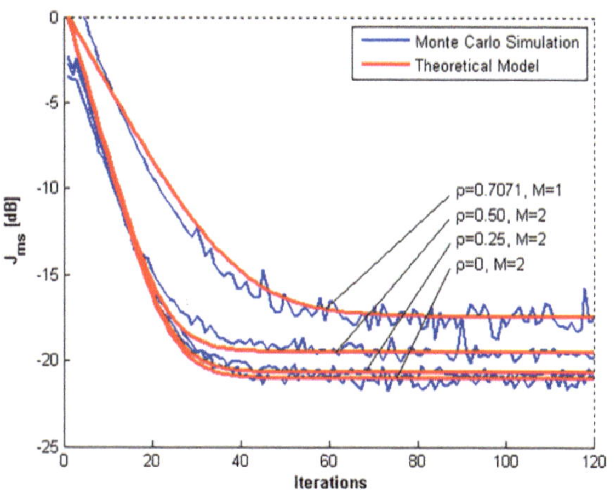

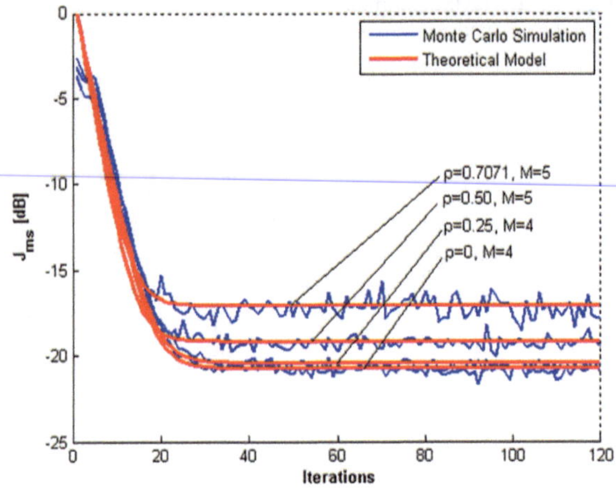


**Fig. 9.5** CKLMS, complexified kernel ($\xi = 0.25$, various $\rho$)

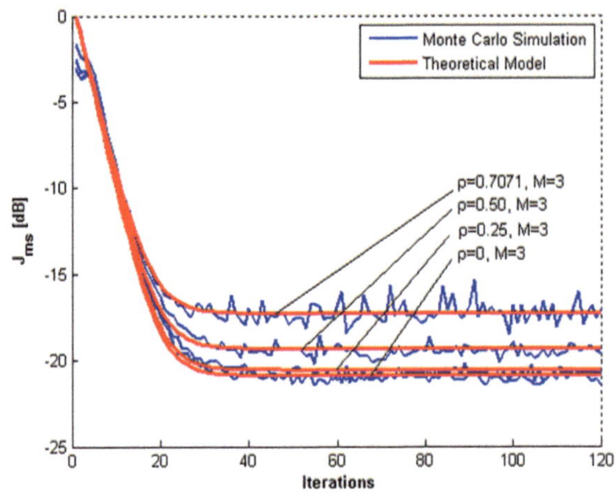


**Fig. 9.6** CKLMS, complexified kernel ($\xi = 0.30$, various $\rho$)

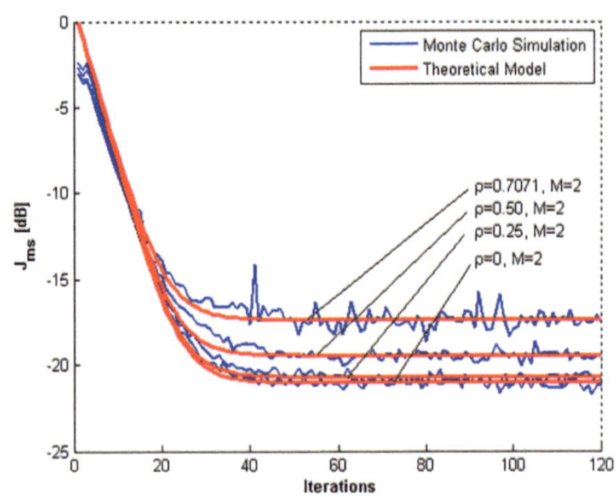


**Fig. 9.7** CKLMS, complexified kernel ($\xi = 0.35$, various $\rho$)

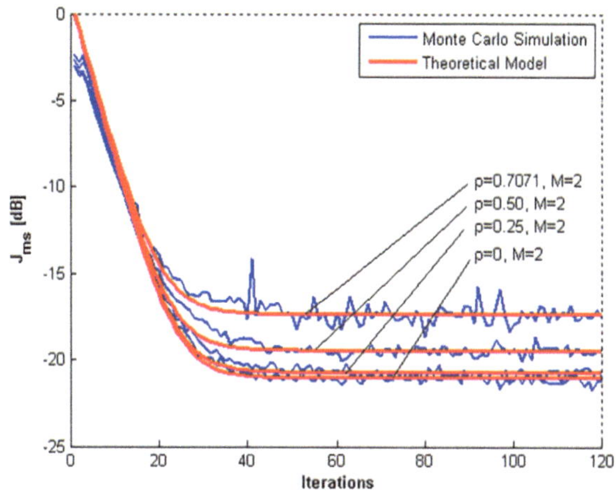


**Fig. 9.8** CKLMS, complexified kernel ($\xi$ = 0.40, various $\rho$)

The value of parameter $\eta$ used for the tests was chosen based on a similar design criteria as [1] (i.e., selected to meet a desired steady-state $J_{ms}(n)$, while ensuring stability based on eigenvalues of $\mathbf{C_v}(n)$).

In each case, the theoretical model and simulations match well, as seen from Figs. 9.1, 9.2, 9.3, 9.4, 9.5, 9.6, 9.7 and 9.8. Figures 9.1, 9.2, 9.3 and 9.4 show the results based on the complex Gaussian kernel (along with Table 9.1), while Figs. 9.5, 9.6, 9.7 and 9.8 and Table 9.2 show the results based on the complexified Gaussian kernel. For both Tables 9.1 and 9.2, $J_{ms\_\infty}$ indicates the value of $J_{ms}(n)$ in steady-state, and $n_\epsilon$ is number of iterations for $J_{ms}(n)$ to reach steady-state, computed with the formula: $\|\mathbf{C_v}(n) - \mathbf{C_v}(n_\epsilon)\| < \epsilon$, where $\epsilon$ is a threshold parameter (chosen to be $\epsilon = 10^{-3}$ here). The value of $J_{\min}$ was evaluated based on estimates of $\mathbf{p_{k}}_d$, $E[|d(n)|^2]$ obtained from averaging the signals for the 500 realizations.

Comparing the results using the complex and complexified kernels, we see that both provide similar minimum MSE performance for each parameter tested (within 0.2 dB in all cases). However, the use of the complexified kernel showed faster convergence rate than the complex kernel, as well as consistency in convergence behavior (i.e., $19 < n_\epsilon < 36$ for the complexified, compared to $23 < n_\epsilon < 176$ for the complex) for parameters tested. Additional testing (i.e., combinations of $\eta$, $\varepsilon_0$ along with $\xi$ and $\rho$) should be considered for full comparison.

It can also be seen that the convergence rate and MSE performance for CKLMS vary with circularity for both the complex and complexified kernel forms, and follow the description in Sect. 9.3. The change in both convergence and MSE performance ($J_{\min}$) across circularity may be seen to be dependent on the channel characteristics (specifically, whether channel behavior is symmetric across real and imaginary input components). The minimum MSE, $J_{\min}$, is observed to increase with circularity for both kernels, indicating non-symmetric behavior for the channel. Similar steady-state MSE was seen using both types of kernels.

**Table 9.1** Settings, results for CKLMS with complex kernel (Figs. 9.1, 9.2, 9.3 and 9.4)

| $\eta$ | $\xi$ | $\rho$ | $M$ | $J_{min}$ (dB) | $J_{ms\_\infty}$ (dB) | $n_{\in}$ |
|---|---|---|---|---|---|---|
| 0.072 | 0.25 | 0 | 4 | −21.119 | −20.780 | 23 |
| 0.072 | 0.30 | 0 | 3 | −21.165 | −20.870 | 25 |
| 0.072 | 0.35 | 0 | 2 | −21.224 | −21.007 | 34 |
| 0.072 | 0.40 | 0 | 2 | −21.262 | −21.028 | 31 |
| 0.072 | 0.25 | 0.25 | 4 | −20.758 | −20.248 | 29 |
| 0.072 | 0.30 | 0.25 | 3 | −20.844 | −20.555 | 26 |
| 0.072 | 0.35 | 0.25 | 2 | −20.882 | −20.667 | 34 |
| 0.072 | 0.40 | 0.25 | 2 | −20.910 | −20.678 | 31 |
| 0.072 | 0.25 | 0.50 | 3 | −19.559 | −19.315 | 122 |
| 0.072 | 0.30 | 0.50 | 2 | −19.637 | −19.448 | 44 |
| 0.072 | 0.35 | 0.50 | 2 | −19.658 | −19.446 | 35 |
| 0.072 | 0.40 | 0.50 | 2 | −19.719 | −19.488 | 31 |
| 0.072 | 0.25 | $1/\sqrt{2}$ | 2 | −17.572 | −17.275 | 172 |
| 0.072 | 0.30 | $1/\sqrt{2}$ | 2 | −17.629 | −17.426 | 76 |
| 0.072 | 0.35 | $1/\sqrt{2}$ | 2 | −17.632 | −17.412 | 35 |
| 0.072 | 0.40 | $1/\sqrt{2}$ | 1 | −17.554 | −17.438 | 67 |

**Table 9.2** Settings, results for CKLMS with complexified kernel (Figs. 9.5, 9.6, 9.7 and 9.8)

| $\eta$ | $\xi$ | $\rho$ | $M$ | $J_{min}$ (dB) | $J_{ms\_\infty}$ (dB) | $n_{\in}$ |
|---|---|---|---|---|---|---|
| 0.072 | 0.25 | 0 | 4 | −21.110 | −20.772 | 23 |
| 0.072 | 0.30 | 0 | 3 | −21.171 | −20.877 | 25 |
| 0.072 | 0.35 | 0 | 2 | −21.238 | −21.022 | 34 |
| 0.072 | 0.40 | 0 | 2 | −21.272 | −21.038 | 31 |
| 0.072 | 0.25 | 0.25 | 4 | −20.772 | −20.444 | 23 |
| 0.072 | 0.30 | 0.25 | 3 | −20.838 | −20.550 | 25 |
| 0.072 | 0.35 | 0.25 | 2 | −20.905 | −20.691 | 34 |
| 0.072 | 0.40 | 0.25 | 2 | −20.943 | −20.711 | 31 |
| 0.072 | 0.25 | 0.50 | 5 | −19.550 | −19.162 | 18 |
| 0.072 | 0.30 | 0.50 | 3 | −19.606 | −19.328 | 26 |
| 0.072 | 0.35 | 0.50 | 2 | −19.672 | −19.463 | 35 |
| 0.072 | 0.40 | 0.50 | 2 | −19.681 | −19.452 | 31 |
| 0.072 | 0.25 | $1/\sqrt{2}$ | 5 | −17.447 | −17.071 | 19 |
| 0.072 | 0.30 | $1/\sqrt{2}$ | 3 | −17.547 | −17.274 | 26 |
| 0.072 | 0.35 | $1/\sqrt{2}$ | 2 | −17.545 | −17.339 | 36 |
| 0.072 | 0.40 | $1/\sqrt{2}$ | 2 | −17.584 | −17.357 | 32 |

**Example 2** In Figs. 9.1, 9.2, 9.3, 9.4, 9.5, 9.6, 9.7 and 9.8, step-size $\eta$ was held constant (for isolating effect of circularity on convergence). Figures 9.9 and 9.10 illustrate different step-size $\eta$, as well as circularity larger than $1/\sqrt{2}$ ($\rho = 0.9$). Otherwise the settings used are the same as in Figs. 9.1, 9.2, 9.3, 9.4, 9.5, 9.6, 9.7

and 9.8, aside from 180 iterations being displayed. The test results and parameters are summarized in Tables 9.3 and 9.4.

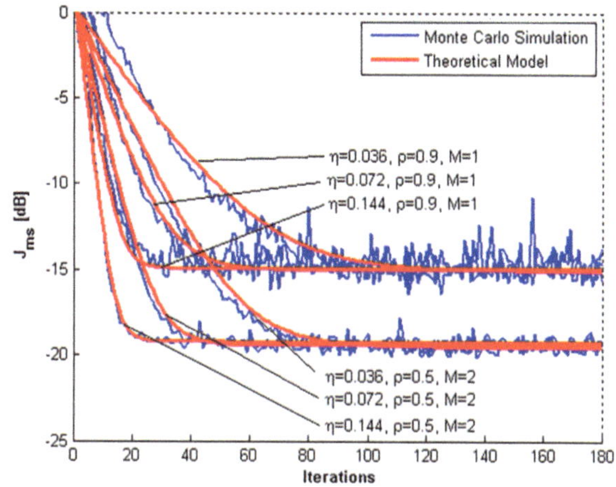


**Fig. 9.9** Complex kernel: $\xi = 0.3,\ \rho \in \{0.5, 0.9\}$

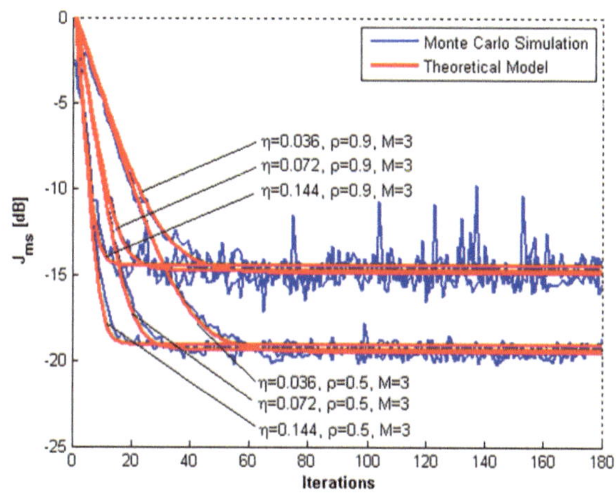


**Fig. 9.10** Complexified: $\xi = 0.3,\ \rho \in \{0.5, 0.9\}$

**Table 9.3** CKLMS settings, results with complex kernel (Fig. 9.9)

| $\eta$ | $\xi$ | $\rho$ | $M$ | $J_{min}$ (dB) | $J_{ms\_\infty}$ (dB) | $n_\in$ |
|---|---|---|---|---|---|---|
| 0.036 | 0.30 | 0.50 | 2 | −19.634 | −19.540 | 84 |
| 0.072 | 0.30 | 0.50 | 2 | −19.637 | −19.448 | 44 |
| 0.144 | 0.30 | 0.50 | 2 | −19.637 | −19.247 | 33 |
| 0.036 | 0.30 | 0.90 | 1 | −15.147 | −15.088 | 128 |
| 0.072 | 0.30 | 0.90 | 1 | −15.142 | −15.026 | 63 |
| 0.144 | 0.30 | 0.90 | 1 | −15.142 | −14.916 | 30 |

**Table 9.4** CKLMS settings, results with complexified kernel (Fig. 9.10)

| $\eta$ | $\xi$ | $\rho$ | $M$ | $J_{min}$ (dB) | $J_{ms\_\infty}$ (dB) | $n_\in$ |
|---|---|---|---|---|---|---|
| 0.036 | 0.30 | 0.50 | 3 | −19.602 | −19.466 | 52 |
| 0.072 | 0.30 | 0.50 | 3 | −19.606 | −19.328 | 26 |
| 0.144 | 0.30 | 0.50 | 3 | −19.606 | −19.029 | 12 |
| 0.036 | 0.30 | 0.90 | 3 | −15.014 | −14.876 | 52 |
| 0.072 | 0.30 | 0.90 | 3 | −15.003 | −14.723 | 29 |
| 0.144 | 0.30 | 0.90 | 3 | −15.001 | −14.418 | 14 |

From these graphs, convergence rate increases with $\eta$, as expected. This applies for both the complex and complexified kernels. The tradeoff between misadjustment and convergence that exists for LMS-based methods is also observed here.

With circularity $\rho = 0.9$ the MSE increases, which can be understood from the description of Example 1. The complex kernel provided an improved MSE (also reduced computation), although convergence is slower. Note that the choice of complex or complexified kernel may depend on the application goal (i.e., emphasize MSE, convergence rate), in addition to the channel.

**Example 3** Finally, Figs. 9.11 and 9.12 illustrate a case using a different nonlinear channel [1], and larger resulting dictionary size. The test methodology is the same as before (i.e., 500 samples are used to build the dictionary). The CKLMS is again run for 1000 iterations, however 700 iterations are now shown.

The linear portion of the channel was

$$y(n) = [10.5] \cdot \boldsymbol{u}(n) - 0.2y(n-1) + 0.35y(n-2) \tag{9.81}$$

and the nonlinearity was

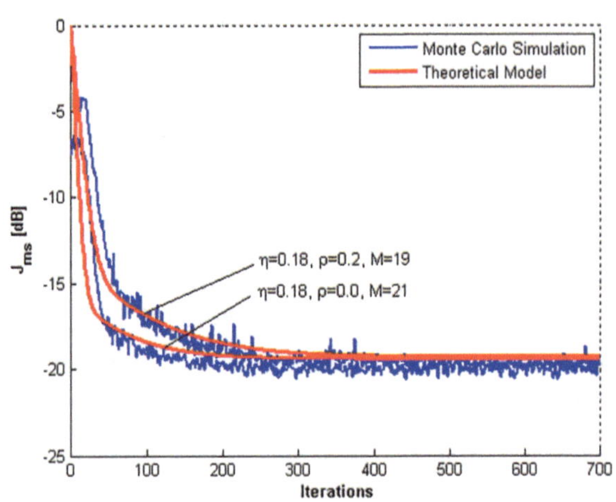


**Fig. 9.11** Complex kernel, channel (9.81) and (9.82)

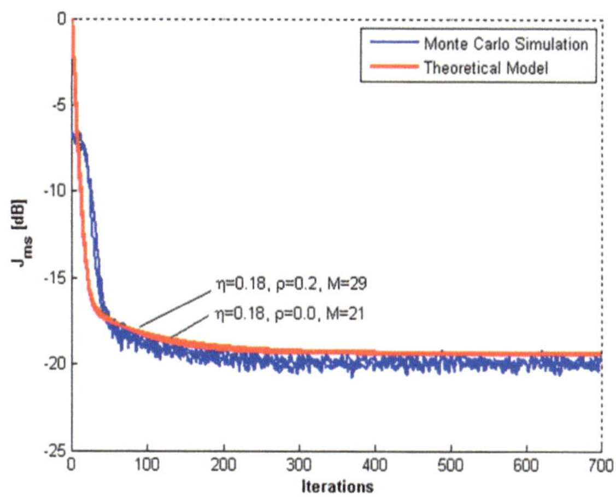


**Fig. 9.12** Complexified, channel (9.81) and (9.82)

$$d(n) = \frac{0.3163 \cdot y(n)}{\sqrt{0.1 + 0.9y^2(n)}} + z(n) \tag{9.82}$$

where $z(n)$ is zero-mean Gaussian with variance ${\sigma_z}^2 = 10^{-4}$.

The input sequence $\mathbf{u}(n)$ used for these tests followed the same statistical description provided after Eq. (9.78), except ${\sigma_{x1}}^2 = {\sigma_{x2}}^2 = 0.0625$. The coherence parameter used was $\varepsilon_0 = 3 \times 10^{-1}$, and the threshold parameter $\epsilon$ for finding the convergence measure $n_\epsilon$ was $\epsilon = 10^{-3}$. All other parameters used are shown in Tables 9.5 and 9.6.

For these graphs also, the theory-predicted result matched simulation, though some discrepancy was observed. This may be due to independence assumptions (see Sect. 9.2). The dictionary length was between 19 and 29 vectors, and slower convergence was observed than for previous tests. The reduced convergence rate is expected since substantially increasing the input vector length (i.e., $\boldsymbol{\kappa}_\omega(n)$ in (9.14)) is known to yield slower convergence for adaptive filters. Convergence is also affected by many factors, however (i.e., $\eta$,$\rho$,$\xi$, as well as the channel used). The complexified

**Table 9.5** CKLMS settings, results with complex kernel (Fig. 9.11)

| $\rho$ | $\xi$ | $M$ | $\eta$ | $J_{min}$ (dB) | $J_{ms\_\infty}$ (dB) | $n_\epsilon$ |
|---|---|---|---|---|---|---|
| 0 | 0.14 | 21 | 0.180 | −20.182 | −19.382 | 242 |
| 0.2 | 0.14 | 19 | 0.180 | −20.090 | −19.279 | 380 |

**Table 9.6** CKLMS settings, results with complexified kernel (Fig. 9.12)

| $\rho$ | $\xi$ | $M$ | $\eta$ | $J_{min}$ (dB) | $J_{ms\_\infty}$ (dB) | $n_\epsilon$ |
|---|---|---|---|---|---|---|
| 0 | 0.14 | 21 | 0.180 | −20.171 | −19.394 | 242 |
| 0.2 | 0.14 | 29 | 0.180 | −20.095 | −19.352 | 329 |

kernel is still observed to provide marginally faster convergence in this case as well, with similar misadjustment.

Examples 1–3 show the theory-predicted expressions match simulations well also. The convergence rate is affected by $\eta$ in the expected manner. Specifically, a large $\eta$ increases the rate of convergence, although misadjustment increases slightly. Smaller values of $\eta$ yield slower convergence and improved misadjustment. The circularity value $\rho = 0.9$ increased the MSE considerably (around 5 dB) compared to $\rho = 0.5$ for both the complex and complexified kernel forms. This result also indicated a benefit for using the complexified kernel on convergence, which appears consistent across circularity in the tests here.

## 9.5 Conclusion

This chapter described a convergence analysis of the complex KLMS (CKLMS). The analysis assumed use of the complex or complexified Gaussian kernel, in addition to using a complex Gaussian random noise input with configurable circularity.

The theoretical model was able to provide predicted MSE results which were seen to match simulations well. Various input circularities were tested (from highly non-circular to circular), which illustrated that the MSE performance and convergence rate varied with circularity.

The analysis of CKLMS performed enables a study of the convergence behavior of the algorithm based on input circularity, in addition to the kernel bandwidth and step-size $\eta$ parameter.

## References

1. W.D. Parreira, J.C.M. Bermudez, C. Richard, J.-Y. Tourneret, Stochastic behavior analysis of the Gaussian Kernel-least-mean-square algorithm. IEEE Trans. Signal Process. **60**(5), 2208–2222 (2012)
2. W.D. Parreira, J.C.M. Bermudez, C. Richard, J.-Y. Tourneret, Stochastic behavior analysis of the Gaussian Kernel least mean square algorithm, in The *IEEE Int. Conference on Acoustics, Speech, and Signal Processing*, pp. 4116–4119 (2011)
3. P. Bouboulis, S. Theodoridis, Extension of Wirtinger's calculus to reproducing Kernel Hilbert spaces and the complex Kernel LMS. IEEE Trans. Signal Process. **59**(3) (2011)
4. P. Bouboulis, S. Theodoridis, The complex Gaussian kernel LMS algorithm, in *Proc. of Int. Conf. on Artificial Neural Networks* (2010)
5. W. Liu, J. Principe, S. Haykin, *Kernel Adaptive Filtering: A Comprehensive Introduction* (Wiley Publishers, 2010)
6. C. Richard, J.C.M. Bermudez, P. Honeine, Online prediction of time series data with Kernels. IEEE Trans. Signal Process. **57**(3), 1058–1066 (2009)

# Chapter 10
# The Quaternion Kernel LMS Algorithm

In this chapter, we describe a kernel adaptive filter for quaternion data, which we refer to as the Quaternion Kernel Least Mean Square (Quat-KLMS) algorithm. A key portion of the derivation involves understanding the design of kernel functions suitable for quaternion data, and also their associated quaternion reproducing kernel Hilbert Space (RKHS). Also, a modified $\mathbb{HR}$ calculus for Hilbert spaces was needed for finding gradients of cost functions defined on a quaternion RKHS.

There are previous references describing the use of quaternion-valued data with kernel methods. For example, Corrochano and Daniel [1] provides an approach for designing support vector machines (SVMs) using Clifford algebra (i.e. an algebra for hyper-complex data). Reference [2] shows Mercer's theorem for quaternionic kernels, which is used to design a support vector regressor for quaternion data in [3]. The method described here is a kernel method based on LMS, providing an online approach whose derivation is simplified compared to existing methods. The simplicity is owed to the gradient of [4], extended here for quaternion Hilbert spaces, as well as the transform mapping interpretation. The benefit of quaternion widely linear estimation is also considered, as well as the i-gradient of Chap. 5.

## 10.1 Combining Quaternions with Kernel Methods

This section describes definitions for quaternion kernels, combining the benefits of quaternions with the power of kernel-based learning. Definitions for complex-valued kernels from [5] serve as a guide, along with [2] to extend for quaternions. Additional background on kernel functions, RKHSs, and their use in nonlinear filtering can be found in [5, 6].

Initially, define a function $\kappa : \boldsymbol{X} \times \boldsymbol{X} \rightarrow \mathbb{H}$, where $x_1, \ldots, x_N \in \boldsymbol{X}$ defines the set of possible inputs, and $\mathbb{H}$ denotes a quaternion division algebra. A division algebra can



T. K. Paul and T. Ogunfunmi, *Adaptive Filter Algorithms For Complex and Quaternion Data*, https://doi.org/10.1007/978-3-032-23798-9_10

be defined as an algebra where division can be performed. Specifically, each non-zero element $q$ of the algebra has a unique left, right inverse element. If, for every non-zero element, the left and right inverse element are the same, the algebra is also considered associative. The quaternion division algebra is an associative algebra, even though multiplication in the algebra is non-commutative in general [4].

It should be noted that the non-commutative property for multiplication can be observed to complicate the formation of kernel methods using quaternions. Nevertheless, we shall proceed to consider a quaternion kernel definition here.

Based on function $\kappa$ from before, define an $N \times N$ matrix $\mathbf{K}$ with elements $\mathbf{K}_{ij} = \kappa\left(x_i, x_j\right)$ for $i, j = 1, \ldots, N$. For $\kappa$ to be a quaternion kernel function, $\mathbf{K}$ should satisfy

$$\mathbf{q}^H \mathbf{K} \mathbf{q} = \sum_{i=1,j=1}^{N,N} q_i^* \mathbf{K}_{i,j} q_j \geq 0 \tag{10.1}$$

for all $q_i \in \mathbb{H}$, $i = 1 \ldots N$, and should also be Hermitian (i.e. $\mathbf{K}^H = \mathbf{K}$). This defines a positive semidefinite matrix in the quaternion case, similar to the complex case. A function $\kappa$ satisfying this may be referred to as a non-negative definite kernel or, for notation convenience, simply as a kernel.

It is well established that non-negative definite kernels have an associated reproducing kernel Hilbert space (RKHS) [3, 5, 6]. In particular, a non-negative definite kernel $\kappa$ can be seen, based on the Mercer theorem, to satisfy the property

$$\kappa(\mathbf{u}, \mathbf{v}) = \sum_{i=0}^{\infty} \lambda_i^{-1} \phi_i(\mathbf{u}) \phi_i^*(\mathbf{v}) \tag{10.2}$$

where $\lambda_i$ are the eigenvalues and $\phi_i$ the eigenfunctions.

For quaternions, in [2] it was verified when $\kappa$ is Hermitian and non-negative definite, the associated eigenvalues can be seen to be real and non-negative. This allows the Mercer theorem to apply, in the quaternion case, with the form of (10.2), and enables the following factoring. If mapping $\boldsymbol{\varphi}(\mathbf{x}) : \mathbf{X} \rightarrow \mathrm{Q}$, $\boldsymbol{\varphi}(\mathbf{x}) = [\lambda_1^{-1/2} \phi_1^*(\mathbf{x}),\ \lambda_2^{-1/2} \phi_2^*(\mathbf{x}), \lambda_3^{-1/2} \phi_3^*(\mathbf{x}) \ldots]$ is constructed, we can write

$$\kappa(\mathbf{u}, \mathbf{v}) = \boldsymbol{\varphi}^H(\mathbf{v}) \boldsymbol{\varphi}(\mathbf{u})$$

where the dimensionality of Q is found from the number of non-zero values of $\lambda_i^{-1}$. Q can be seen to correspond to the RKHS of kernel $\kappa$ [6], in this case a quaternion RKHS. Note $\kappa$ defines an inner product in the RKHS, or

$$\kappa(\mathbf{u}, \mathbf{v}) = \langle \widehat{\boldsymbol{\Phi}}(\mathbf{u}), \widehat{\boldsymbol{\Phi}}(\mathbf{v}) \rangle_{\mathrm{Q}}$$

where notation $\widehat{\boldsymbol{\Phi}}_{\mathrm{Q}}(\cdot)$ is referred to as a kernel mapping of the input into quaternion RKHS Q, and $< \cdot, \cdot >_{\mathrm{Q}}$ is its inner product.

The characteristic that the associated eigenvalues are real and nonnegative is based on assuming a Hermitian-based symmetry for quaternion kernels (see [2]). The question may exist whether positive semi-definite quaternion matrices may be formed with a less restricted eigenvalue decomposition, defining alternative kernel functions. However, here we only consider quaternion kernels possessing this symmetry.

Based on these definitions, we now state some quaternion kernels based on the Gaussian kernel form. Derivations of both are outlined in the next section.

A kernel form working directly on quaternion data, referred to here as the *quaternion Gaussian* kernel, we define as

$$\kappa_{\sigma,Q^d}(\mathbf{x},\mathbf{w}) = \exp\left(-\sigma^2 \sum_{j=1}^{d} (x_j - w_j^*)^2\right) \tag{10.3}$$

where $\mathbf{x}$, $\mathbf{w}$ are the input vectors, both of length $d$, consisting of quaternion components. Its associated RKHS is denoted as $Q^d$, and its dimensionality varies based on $d$.

Meanwhile, the kernel function resulting from extending the real-valued Gaussian kernel to quaternion data (similar to the complexification approach of [5]) was seen to be

$$\kappa_{\sigma,\mathbb{Q}^d}(\mathbf{x},\ \mathbf{w}) = 4\exp\left(-\sigma^2 \sum_{j=1}^{d} \left|x_j - w_j\right|^2\right) \tag{10.4}$$

We refer to this approach as the *quaternion-extended real Gaussian* kernel, or Quat-Ext Gaussian kernel for short. Its RKHS is denoted $\mathbb{Q}^d$, where dimensionality varies with $d$.

Using these functions, we will later consider various kernel-based learning methods for quaternions, both linear as well as widely linear approaches (Sect. 10.4).

## 10.2 Reproducing Kernel Hilbert Spaces for Quaternions

In this section, we describe the definition of the quaternion kernels mentioned, as well as their associated reproducing kernel Hilbert spaces. This section also provides insight into how a quaternion RKHS may be formed, despite the non-commutative nature of quaternion multiplication. It also allows for understanding kernel-based learning using these spaces.

### 10.2.1 Quaternion Gaussian Kernel

For a Gaussian kernel function applicable to quaternions, we initially consider the real and complex Gaussian kernel forms. The goal is to provide a quaternion kernel definition with a similar RKHS.

The real Gaussian kernel may be defined [5, 6]

$$\kappa_{\sigma,\mathcal{R}^d}(\mathbf{a},\mathbf{b}) = \exp\left(-\sigma^2 \sum_{i=1}^{d} (a_i - b_i)^2\right) \tag{10.5}$$

where **a**, **b** are real-valued input vectors of length $d$, and $\mathcal{R}^d$ denotes the associated RKHS (dimensionality affected by $d$).

Meanwhile, the complex Gaussian kernel is [5, 8]

$$\kappa_{\sigma,\mathcal{C}^d}(\mathbf{a},\mathbf{b}) = \exp\left(-\sigma^2 \sum_{i=1}^{d} (a_i - b_i^*)^2\right) \tag{10.6}$$

where **a**, **b** are now complex-valued vectors, and $\mathcal{C}^d$ denotes the associated complex RKHS.

The RKHS of the complex Gaussian kernel is considered in detail in [8], where its associated RKHS is seen to match that of the real kernel. In particular, when the input is restricted to real values, the kernel functions are equivalent. However, (10.6) provides a Hermitian inner product definition and is therefore suitable for complex data.

Considering this, for the quaternion Gaussian kernel, we similarly define

$$\kappa_{\sigma,\mathcal{Q}^d}(\mathbf{a},\mathbf{b}) = \exp\left(-\sigma^2 \sum_{i=1}^{d} (a_i - b_i^*)^2\right) \tag{10.7}$$

where vectors **a**, **b** consist of quaternion-valued components, and $\mathcal{Q}^d$ denotes the associated quaternion RKHS.

To verify the RKHS of the quaternion Gaussian kernel, a concern is that multiplication with quaternions, unlike real and complex data, is non-commutative.

Therefore, establishing the equivalence of (10.7) to the inner product $\kappa(\mathbf{a},\mathbf{b}) = \boldsymbol{\varphi}^H(\mathbf{a})\boldsymbol{\varphi}(\mathbf{b})$ where $\boldsymbol{\varphi}(\cdot)$ is the associated mapping to RKHS, may be seen to be nontrivial. This will be considered here.

We start by considering the quaternion exponential function, which may be expressed [9] as

$$\exp(q) = \exp(q_a)\big(\cos(\alpha) + \sin(\alpha)\,\alpha^{-1}(iq_b + jq_c + kq_d)\big) \tag{10.8}$$

where $\alpha = \sqrt{q_b^2 + q_c^2 + q_d^2}$, and the functions exp(·), cos(·) and sin(·) on the right side follow the definitions of the real-valued case.

Expanding (10.8) using the definitions of these functions:

$$\exp(q) = \left(\sum_{n=0}^{\infty} \frac{q_a^n}{n!}\right)\left(\sum_{m=0}^{\infty} \frac{(-1)^m \alpha^{2m}}{(2m)!} + \sum_{m=0}^{\infty} \frac{(-1)^m \alpha^{2m}}{(2m+1)!}(iq_b + jq_c + kq_d)\right)$$
$$= \left(\sum_{n=0}^{\infty} \frac{q_a^n}{n!}\right)\left(\sum_{m=0}^{\infty} (-1)^m \alpha^{2m}\left[(2m)!^{-1} + (2m+1)!^{-1}(iq_b + jq_c + kq_d)\right]\right) \tag{10.9}$$

and using the properties of quaternions, we get

$$\exp(q) = \left(\sum_{n=0}^{\infty} n!^{-1}(\tfrac{1}{2})^n (q+q^*)^n\right)$$
$$\times \left(\sum_{m=0}^{\infty} (\tfrac{1}{2})^{2m}(q-q^*)^{2m}\left[(2m)!^{-1} + (2m+1)!^{-1}\tfrac{1}{2}(q-q^*)\right]\right)$$
$$= \sum_{n=0}^{\infty}\sum_{m=0}^{\infty}\left[\frac{(\frac{1}{2})^{n+2m}}{n!(2m)!}(q+q^*)^n(q-q^*)^{2m} + \frac{(\frac{1}{2})^{n+2m+1}}{n!(2m+1)!}(q+q^*)^n(q-q^*)^{2m+1}\right] \tag{10.10}$$

Finally, using (10.10) with (10.7) yields the following. Here we consider the scalar input case only ($d = 1$) for tractability:

$$\kappa_{\sigma,\mathbb{Q}^d}(a,b)$$
$$= \sum_{n=0}^{\infty}\sum_{m=0}^{\infty} [\frac{(-\sigma^2/2)^{n+2m}}{n!(2m)!}((a-b^*)^2 + (a^*-b)^2)^n((a-b^*)^2 - (a^*-b)^2)^{2m}$$
$$+ \frac{(-\sigma^2/2)^{n+2m+1}}{n!(2m+1)!}((a-b^*)^2 + (a^*-b)^2)^n(a-b^*)^2 - (a^*-b)^2)^{2m+1}] \tag{10.11}$$

The vector input case may be considered as with [8].

Expanding and factoring of (10.11) is required to yield an inner product form. It is observed that, even for the scalar input case, this process is non-trivial.

Instead, we initially move our attention to alternative kernel functions based on the WL method of Chap. 5 (these kernel forms are outlined later in Sects. 10.4.2 and 10.4.3). These can be expressed as

$$\kappa_{\mathbb{Q}_{WL}}(\mathbf{a},\mathbf{b}) = \kappa_{\mathbb{Q}}(\mathbf{a},\mathbf{b}) + \kappa_{\mathbb{Q}}^i(\mathbf{a},\mathbf{b}) + \kappa_{\mathbb{Q}}^j(\mathbf{a},\mathbf{b}) + \kappa_{\mathbb{Q}}^k(\mathbf{a},\mathbf{b})$$

$$= 4\mathrm{Re}\left\{\exp\left(-\sigma^2\sum_{i=1}^{d}(a_i - b_i^*)^2\right)\right\} \tag{10.12a}$$

$$\begin{aligned}\kappa_{Q_{IB}}(\mathbf{a}, \mathbf{b}) &= \kappa_Q\left(\left[\mathbf{a}, \mathbf{a}^i, \mathbf{a}^j, \mathbf{a}^k\right], \left[\mathbf{b}, \mathbf{b}^i, \mathbf{b}^j, \mathbf{b}^k\right]\right)\\ &= \exp\left(-\sigma^2\sum_{i=1}^{d}4\mathrm{Re}\left\{(a_i - b_i^*)^2\right\}\right)\end{aligned} \tag{10.12b}$$

where (10.7) is used for $\kappa_Q(\cdot, \cdot)$.

For analyzing $K_{Q_{WL}}(\mathbf{a}, \mathbf{b})$, considering the quaternion exponential definition (10.8), we can write

$$\kappa_{Q_{WL}}(\mathbf{a}, \mathbf{b}) = 4\exp\left(-\sigma^2\sum_{i=1}^{d}\mathrm{Re}\left\{(a_i - b_i^*)^2\right\}\right)\cos\left(\sigma^2\left|\sum_{i=1}^{d}\mathrm{Im}\left\{(a_i - b_i^*)^2\right\}\right|\right) \tag{10.13}$$

where exp($\cdot$) and cos($\cdot$) use the real-valued definitions. In a similar fashion to (10.11), the tractability of (10.13) seems poor.

However, for $K_{Q_{IB}}(\mathbf{a}, \mathbf{b})$, since this can be expressed as

$$\begin{aligned}\kappa_{Q_{IB}}(\mathbf{a}, \mathbf{b}) &= \exp\left(-\sigma^2\sum_{l=1}^{d}(a_l - b_l^*)^2 + \left[(a_l - b_l^*)^2\right]^i\right.\\ &\quad\left. + \left[(a_l - b_l^*)^2\right]^j + \left[(a_l - b_l^*)^2\right]^k\right)\\ &= \exp\left(-\sigma^2\sum_{l=1}^{d}4\mathrm{Re}\left\{(a_l - b_l^*)^2\right\}\right)\end{aligned}$$

we can evaluate (10.12b) directly using the real-valued definition of exp($\cdot$). Here we consider the $d = 1$ case, yielding

$$\begin{aligned}\kappa_{Q_{IB}}(a, b) &= \exp\left(-\sigma^2 4\mathrm{Re}\left\{(a - b^*)^2\right\}\right)\\ &= \exp(4\sigma^2[-\mathrm{Re}\{a^2\} + \mathrm{Re}\{b^*a + ab^*\} - \mathrm{Re}\{b^{*2}\}])\\ &= \exp(-4\sigma^2\mathrm{Re}\{a^2\})\exp(4\sigma^2\mathrm{Re}\{b^*a + ab^*\})\exp(-4\sigma^2\mathrm{Re}\{b^{*2}\})\\ &= \exp(-4\sigma^2\mathrm{Re}\{a^2\})\left[\sum_{n=0}^{\infty}\frac{(4\sigma^2\mathrm{Re}\{b^*a + ab^*\})^n}{n!}\right]\exp(-4\sigma^2\mathrm{Re}\{b^{*2}\})\end{aligned} \tag{10.14}$$

Evaluating the summation term in (10.14) using the properties of Chap. 5 yields

$$
\begin{aligned}
&\sum_{n=0}^{\infty} \frac{\left(4\sigma^2 \mathrm{Re}\{ab^* + b^*a\}\right)^n}{n!} = \sum_{n=0}^{\infty} \frac{\left(2\sigma^2 [ab^* + b^*a + ba^* + a^*b]\right)^n}{n!} \\
&= \sum_{n=0}^{\infty} \frac{\left(2\sigma^2 [ab^* + a^*b + \frac{1}{2}(a^{*i}b^i + a^{*j}b^j + a^{*k}b^k - a^*b) + \frac{1}{2}(a^i b^{*i} + a^j b^{*j} + a^k b^{*k} - ab^*)]\right)^n}{n!} \\
&= \sum_{n=0}^{\infty} \frac{\left(\sigma^2 [ab^* + a^i b^{*i} + a^j b^{*j} + a^k b^{*k} + a^*b + a^{*i}b^i + a^{*j}b^j + a^{*k}b^k]\right)^n}{n!}
\end{aligned}
\tag{10.15}
$$

Note that the $n = 1$ term of (10.15) can be readily expressed in an inner product form. Specifically

$$
\begin{aligned}
&\sigma^2(a^*b + a^{*i}b^i + a^{*j}b^j + a^{*k}b^k + ab^* + a^i b^{*i} + a^j b^{*j} + a^k b^{*k}) \\
&= \boldsymbol{\psi}_1^H(a)\boldsymbol{\psi}_1(b)
\end{aligned}
\tag{10.16}
$$

where $\boldsymbol{\psi}_1(x) = \sigma\left[x, x^i, x^j, x^k, x^*, x^{*i}, x^{*j}, x^{*k}\right]^T$. Therefore, it remains to be seen if the $n > 1$ terms may be expressed with inner products.

This analysis is performed in Appendix E, which attempts to show that these terms may be expanded as

$$
\begin{aligned}
&(\sigma^{2n}/n!)(a^*b + a^{*i}b^i + a^{*j}b^j + a^{*k}b^k + ab^* + a^i b^{*i} + a^j b^{*j} + a^k b^{*k})^n \\
&= \boldsymbol{\psi}_n^H(a)\boldsymbol{\psi}_n(b)
\end{aligned}
\tag{10.17}
$$

where $\boldsymbol{\psi}_n(x) = \left(\sigma^{2n}/n!\right)^{1/2}\left[x^{\cdots}xx, x^{\cdots}xx^i, x^{\cdots}xx^j, \ldots, x^{*k\cdots}x^{*k}x^{*k}\right]^T$, which consists of $8^n$ elements, that are all the permutations of multiplying $n$ elements, chosen with repetition, from the set: $\left\{x, x^i, x^j, x^k, x^*, x^{*_i}, x^{*j}, x^{*k}\right\}$.

And for kernel function (10.14), we could write

$$
\begin{aligned}
\kappa_{\mathbb{Q}_{IB}}(a, b) &= \exp\left(-\sigma^2 4\mathrm{Re}\{(a - b^*)^2\}\right) \\
&= \boldsymbol{\varphi}^H(a)\boldsymbol{\varphi}(b)
\end{aligned}
\tag{10.18}
$$

where $\boldsymbol{\varphi}(x) = \exp(-4\sigma^2\mathrm{Re}\{x^2\})[\boldsymbol{\psi}_0^T(x),\ \boldsymbol{\psi}_1^T(x),\ \boldsymbol{\psi}_2^T(x),\ \boldsymbol{\psi}_3^T(x)\ \ldots\ ]^T$.

An important property for the associated transform mapping shown in (10.18) is $\boldsymbol{\varphi}(x^\eta) = \boldsymbol{\varphi}^\eta(x)$ for $\eta \in \{i, k, j\}$, which can be seen from (10.17), since $a^\eta b^\eta = (ab)^\eta$ for $\eta \in \{i, k, j\}$, and also $(a^\omega)^\rho = (a^\rho)^\omega$ for $\omega, \rho \in \{i, k, j\}$. The property $\boldsymbol{\varphi}(x^\eta) = \boldsymbol{\varphi}^\eta(x)$ should extend to the vector case of $\kappa_{\mathbb{Q}_{IB}}(\mathbf{a}, \mathbf{b})$ of (10.12b). This is also assumed to apply to (10.7), as well as $\kappa_{\mathbb{Q}_{WL}}(\mathbf{a}, \mathbf{b})$ of (10.12a), however, the transform spaces for these functions require verification. These properties were used in deriving the various algorithms obtained in Sect. 10.4.

We now describe an alternative method of obtaining a Gaussian kernel suitable for quaternion inputs.

### 10.2.2 Quaternion-Extended Gaussian Kernel

In order to extend a real-valued kernel function for use with quaternions, we consider an approach similar to the derivation of the complexified real Gaussian kernel from [5].

Initially, let $X \subseteq \mathbb{R}^{\nu}$ and $X^4 \equiv X \times X \times X \times X \subseteq \mathbb{R}^{4\nu}$. $X^4$ is the subspace representing the components of a $\nu$-dimensional vector of quaternions. Then define the quaternion vector space $\mathbb{X} = \{x_a + ix_b + jx_c + kx_d,\ x_a, x_b, x_c, x_d \in X\} \subseteq \mathbb{H}^{\nu}$ with a quaternion product structure. We now define the RKHS $\mathcal{H}$ associated with a real kernel $\kappa$ defined on $X^4 \times X^4$, and its inner product $< , >_{\mathcal{H}}$. Finally, let $\mathcal{H}^4 = \mathcal{H} \times \mathcal{H} \times \mathcal{H} \times \mathcal{H}$. Then $\mathcal{H}^4$ can be seen to be a Hilbert space with inner product:

$$\langle \mathbf{f}, \mathbf{g} \rangle_{\mathcal{H}^4} = \langle f_1, g_1 \rangle_{\mathcal{H}} + \langle f_2, g_2 \rangle_{\mathcal{H}} + \langle f_3, g_3 \rangle_{\mathcal{H}} + \langle f_4, g_4 \rangle_{\mathcal{H}} \tag{10.19}$$

with $f = (f_1, f_2, f_3, f_4)$, $g = (g_1, g_2, g_3, g_4)$. The Hilbert space $\mathcal{H}$ can be seen to represent the RKHS for learning a single component of a quaternion function output, while $\mathcal{H}^4$ is the RKHS for learning all four components. However, we still require a quaternion structure. Define space $\mathbb{Q} = \{f = f_1 + if_2 + jf_3 + kf_4, f_1, f_2, f_3, f_4 \in \mathcal{H}\}$ with quaternion inner product

$$\begin{aligned}\langle f, g \rangle_{\mathbb{Q}} = &\langle f_1, g_1 \rangle_{\mathcal{H}} + \langle f_2, g_2 \rangle_{\mathcal{H}} + \langle f_3, g_3 \rangle_{\mathcal{H}} + \langle f_4, g_4 \rangle_{\mathcal{H}} \\ &+ i\big(\langle f_2, g_1 \rangle_{\mathcal{H}} - \langle f_1, g_2 \rangle_{\mathcal{H}} + \langle f_4, g_3 \rangle_{\mathcal{H}} - \langle f_3, g_4 \rangle_{\mathcal{H}}\big) \\ &+ j\big(\langle f_3, g_1 \rangle_{\mathcal{H}} - \langle f_1, g_3 \rangle_{\mathcal{H}} + \langle f_2, g_4 \rangle_{\mathcal{H}} - \langle f_4, g_2 \rangle_{\mathcal{H}}\big) \\ &+ k\big(\langle f_4, g_1 \rangle_{\mathcal{H}} - \langle f_1, g_4 \rangle_{\mathcal{H}} + \langle f_3, g_2 \rangle_{\mathcal{H}} - \langle f_2, g_3 \rangle_{\mathcal{H}}\big)\end{aligned} \tag{10.20}$$

with $f = f_1 + if_2 + jf_3 + kf_4$, $g = g_1 + ig_2 + jg_3 + kg_4$. Therefore $\boldsymbol{f}, \boldsymbol{g} \in \mathbb{Q}$ are functions defined $\boldsymbol{f}, \boldsymbol{g} : \mathbb{X} \to \mathbb{H}$, where $\mathbb{X} \subseteq \mathbb{H}^{\nu}$. We note that $\mathbb{Q}$ can be seen to be a quaternion RKHS with real kernel $\kappa$ from before. We refer to $\mathbb{Q}$ as a quaternion-extended RKHS.

For mapping quaternion $x$ to RKHS $\mathbb{Q}$, we can use the following rule:

$$\begin{aligned}\widehat{\mathbf{\Phi}}(x) &= \widehat{\mathbf{\Phi}}(x_a + ix_b + jx_c + kx_d) \\ &= \widehat{\mathbf{\Phi}}(x_a, x_b, x_c, x_d) \\ &= \Phi(x_a, x_b, x_c, x_d) + i \cdot \Phi(x_a, x_b, x_c, x_d) + \dots + k \cdot \Phi(x_a, x_b, x_c, x_d)\end{aligned} \tag{10.21}$$

where $\widehat{\mathbf{\Phi}}$ is the mapping applied to quaternion input $x = x_a + ix_b + jx_c + kx_d$, and $\Phi$ is the feature map of real kernel $\kappa$, defined $\Phi(x_a, x_b, x_c, x_d) = \kappa(\cdot, (x_a, x_b, x_c, x_d))$.

With this approach, the mapping to each real component of the quaternion output, based on the real components of the quaternion input, is chosen to be the same. Specifically, feature map $\Phi$ of real kernel $\kappa$ is used. The mapping can be simplified to

$$\widehat{\mathbf{\Phi}}(x) = \Phi(x_a, x_b, x_c, x_d) + i\,\Phi(x_a, x_b, x_c, x_d) + \ldots + k\,\Phi(x_a, x_b, x_c, x_d)$$
$$= (1 + i + j + k) \cdot \Phi(x_a, x_b, x_c, x_d) \tag{10.22}$$

This is due to the fact that the kernel transform mapping $\Phi(x_a, x_b, x_c, x_d)$ is the same for each orthogonal component. Note the mapping associated with a real kernel function is a real-valued RKHS. For the real Gaussian kernel, an exact RKHS description is in [8] and is also summarized in Appendix A.

The inner product of two vectors in RKHS $\mathbb{Q}$ is

$$< \widehat{\mathbf{\Phi}}(\mathbf{x}), \widehat{\mathbf{\Phi}}(\mathbf{w}) >_{\mathbb{Q}} = \widehat{\mathbf{\Phi}}^{H}(\mathbf{w})\widehat{\mathbf{\Phi}}(\mathbf{x})$$
$$= \Phi^T(\mathbf{w}_a, \mathbf{w}_b, \mathbf{w}_c, \mathbf{w}_d)\Phi(\mathbf{x}_a, \mathbf{x}_b, \mathbf{x}_c, \mathbf{x}_d)|1 - i - j - k|^2$$
$$= 4\,\kappa((\mathbf{x}_a, \mathbf{x}_b, \mathbf{x}_c, \mathbf{x}_d), (\mathbf{w}_a, \mathbf{w}_b, \mathbf{w}_c, \mathbf{w}_d))$$

where $\mathbf{w} = \mathrm{w}_a + i\mathrm{w}_b + j\mathrm{w}_c + k\mathrm{w}_d$ and $\mathbf{x} = \mathrm{x}_a + i\mathrm{x}_b + j\mathrm{x}_c + k\mathrm{x}_d$ are quaternion data vectors.

This result, similar to the complexified real kernel trick of [5], shows that any linear method formed from quaternion inner products of form $\mathrm{w}^H\mathrm{x}$ can be converted to a nonlinear method with this approach via replacing each of the inner products using a positive definite real kernel, whose arguments are the real components of the inputs x and w, i.e. $\kappa((\mathbf{x}_a, \mathbf{x}_b, \mathbf{x}_c, \mathbf{x}_d), (\mathbf{w}_a, \mathbf{w}_b, \mathbf{w}_c, \mathbf{w}_d))$. This forms an alternative RKHS for quaternion nonlinear adaptive algorithms.

With the real Gaussian kernel, which is expressed as

$$\kappa_{\sigma,\mathrm{R}^d}(\mathbf{u}, \mathbf{u}') = \exp\left(-\sigma^2 \sum_{j=1}^{d} \left(u_j - u_j'\right)^2\right) \tag{10.23}$$

The quaternion-extended real Gaussian kernel (or Quat-Ext Gaussian kernel for short) is

$$\kappa_{\sigma,\mathbb{Q}^d}(\mathbf{x},\ \mathbf{w}) = 4\exp(-\sigma^2 \sum\nolimits_{j=1}^{d} [(x_{a,j} - w_{a,j})^2 + \ \ldots\ + (x_{d,j} - w_{d,j})^2])$$
$$= 4\exp(-\sigma^2 \sum_{j=1}^{d} \left|x_j - w_j\right|^2) \tag{10.24}$$

The function here satisfies the Mercer condition and may be considered a valid kernel form. It can be seen as a measure of distance between the input vectors and is also symmetric, which are beneficial characteristics of kernels [6].

However, quaternion multiplication is non-commutative. Because of this, the algorithm may not be easily cast in terms of inner products. In this case, we use the transform map $\widehat{\mathbf{\Phi}}(\mathbf{x})$ for the quaternion vector $\mathbf{x}$ directly, which we express as

$$\widehat{\boldsymbol{\Phi}}(\mathbf{x}) = s_{\Phi} \cdot \Phi(\mathbf{x}_{comp}) \tag{10.25}$$

where $s_{\Phi} = (1+i+j+k)$ is the quaternion scalar indicating the mapping direction, and $\mathbf{x}_{\text{comp}} = \left[\mathbf{x}_a^T\ \mathbf{x}_b^T\ \mathbf{x}_c^T\ \mathbf{x}_d^T\right]^T$ is a vector of components of $\mathbf{x}$.

## 10.3 Modified Calculus for Hilbert Spaces

A modified $\mathbb{HR}$ calculus for Hilbert spaces is described here to find the gradient of cost functions defined on the newly formed quaternion RKHS. The modified calculus uses the approach of Chap. 4, and is formed based on the Fréchet derivative, which is applicable to general Hilbert spaces.

We initially consider a Hilbert space $H$ for learning functions whose outputs are defined on the quaternion division algebra $\mathbb{H}$. It is known that to generalize differentiability to Banach or Hilbert spaces, the Fréchet derivative may be used. Note that a Hilbert space is just a Banach space whose norm is evaluated based on an inner product, and Hilbert spaces are of interest to us here. A definition of the Fréchet derivative follows.

Define on Hilbert space $H$ an operator $\boldsymbol{T}$: $H \to \mathbb{H}^{\nu}$. Operator $\boldsymbol{T}$ may be referred to as Fréchet differentiable at $f_0$ if a linear continuous operator $\boldsymbol{W} = (W_1, \ldots, W_{\nu})$: $H \to \mathbb{H}^{\nu}$ exists such that

$$\lim_{\|h\|_H \to 0} = \frac{\|\mathbf{T}(f_0 + h) - \boldsymbol{T}(f_0) - \boldsymbol{W}(h)\|_{\mathbb{H}^{\nu}}}{\|h\|_H} = 0 \tag{10.26}$$

where $\|\cdot\|_H = \sqrt{\langle \cdot\ ,\ \cdot \rangle_H}$ is the induced norm, and $\langle \cdot\ ,\ \cdot \rangle_H$ the dot product, on Hilbert space $H$. $\boldsymbol{W}$ is called the Fréchet derivative, and can be denoted $d\boldsymbol{T}(f_0)$.

This Fréchet derivative definition only differs from [5], for complex data, in that the numerator magnitude is measured on $\mathbb{H}$, and since $H$ is a Hilbert space suitable for quaternions. The various Fréchet derivative cases of (10.26) (for $\nu = 1$, $\nu > 1$, and partial derivatives) are similar to those of Ref. [5]. For clarity, these cases are summarized here.

When $\nu = 1$, $\boldsymbol{T}$ and $\boldsymbol{W}$ are scalar, i.e. $T$: $H \to \mathbb{H}$, $W$: $H \to \mathbb{H}$. In this case, based on Reisz's representation theorem, $W$ has inner product form: $W(h) =< h, w >_H$, where $w^*$ is called the gradient of $T$, denoted $w^* = \nabla T(f_0)$, in Eq. (10.26).

When $\nu > 1$ we have the vector: $\boldsymbol{T} = (T_1, T_2, \ldots, T_{\nu})$, and the Fréchet derivative can be seen to be

$$\partial \boldsymbol{T}(f_0)(h) = \begin{pmatrix} \langle h, \nabla T_1(f_0)^* \rangle_H \\ \vdots \\ \langle h, \nabla T_{\nu}(f_0)^* \rangle_H \end{pmatrix} \tag{10.27}$$

To consider partial derivatives, define $T$: $H^{\mu} \rightarrow \mathbb{H}$ on Hilbert space $H^{\mu}$ equipped with inner product:

$$\langle \boldsymbol{f}, \boldsymbol{g} \rangle_{H^{\mu}} = \sum_{\iota=1}^{\mu} \langle f_{\iota}, g_{\iota} \rangle_{H} \tag{10.28}$$

where $\boldsymbol{f} = (f_1, f_2, \ldots, f_{\mu})$, $\boldsymbol{g} = (g_1, g_2, \ldots, g_{\mu})$. $T(\boldsymbol{f})$ is considered Fréchet differentiable at $\boldsymbol{f}_0$ w.r.t. $f_{\iota}$ iff there exists $w \in H$ so that

$$\lim_{\|h\|_H \to 0} = \frac{T(\boldsymbol{f}_0 + [h]_{\iota}) - T(\boldsymbol{f}_0) - \langle [h]_{\iota}, w \rangle_H}{\|h\|_H} = 0 \tag{10.29}$$

where $[h]_i = (0, 0, \ldots, 0, h, 0, \ldots, 0)$ is the element of $H^{\mu}$ with zero entries everywhere except at place $\iota$, and $w^*$ is the gradient of $T$ at $\boldsymbol{f}_0$ w.r.t. $f_{\iota}$, denoted $w^* = \nabla_{\iota} T(\boldsymbol{f}_0)$.

The Fréchet partial derivative at $\boldsymbol{f}_0$ w.r.t. $f_{\iota}$ can be denoted

$$\frac{\partial T}{\partial f_{\iota}}(\boldsymbol{f}_0)(h) = \langle [h]_{\iota}, w \rangle_H \tag{10.30}$$

These definitions provide the foundations of how the Fréchet derivative is considered.

For describing a modified $\mathbb{HR}$ calculus, we consider the Hilbert space $\mathbb{Q}$ described in Sect. 10.2.2, which was equipped with a suitable quaternion inner product. It may be seen that space $\mathbb{Q}$ is isomorphic with $\mathcal{H}^4$ (or a real RKHS). This isomorphism can be used to describe Fréchet-based differentiation rules with quaternion-valued functions defined on $\mathbb{Q}$, in a similar way to the rules for complex data in [5].

Consider a function $\boldsymbol{T}$ defined on an open subset $A$ of $\mathbb{Q}$, i.e. $\boldsymbol{T} : A \subseteq \mathbb{Q} \rightarrow \mathbb{H}$, where

$$\begin{aligned} \boldsymbol{T}(\boldsymbol{f}) &= \boldsymbol{T}(\boldsymbol{f}_a + i\boldsymbol{f}_b + j\boldsymbol{f}_c + k\boldsymbol{f}_d) \\ &= T_a(\boldsymbol{f}_a, \boldsymbol{f}_b, \boldsymbol{f}_c, \boldsymbol{f}_d) + iT_b(\boldsymbol{f}_a, \boldsymbol{f}_b, \boldsymbol{f}_c, \boldsymbol{f}_d) + \ \ldots \ + kT_d(\boldsymbol{f}_a, \boldsymbol{f}_b, \boldsymbol{f}_c, \boldsymbol{f}_d) \\ &= [T_a(\boldsymbol{f}_a, \boldsymbol{f}_b, \boldsymbol{f}_c, \boldsymbol{f}_d),\ T_b(\boldsymbol{f}_a, \boldsymbol{f}_b, \boldsymbol{f}_c, \boldsymbol{f}_d),\ \ldots,\ T_d(\boldsymbol{f}_a, \boldsymbol{f}_b, \boldsymbol{f}_c, \boldsymbol{f}_d)] \end{aligned} \tag{10.31}$$

with $\boldsymbol{f}_a, \boldsymbol{f}_b, \boldsymbol{f}_c, \boldsymbol{f}_d \in \mathcal{H}$, and $T_a$, $T_b$, $T_c$, $T_d$ being real-valued functions defined on $\mathcal{H}^4$. The second step equivalence is based on the isomorphism between $\mathbb{Q}$ and $\mathcal{H}^4$, and the final step the isomorphism between $\mathbb{H}$ and $\mathbb{R}^4$ of the output.

If $T_a$, $T_b$, $T_c$ and $T_d$ are Fréchet differentiable in the real sense, then we can describe (regardless of whether a Fréchet-based quaternion derivative of $\boldsymbol{T}$ based on (10.26) exists) differentiation rules using the Fréchet real derivative applicable to quaternion functions on $\mathbb{Q}$. The rules here should naturally extend the $\mathbb{HR}$ calculus of [4] to Hilbert spaces.

From the first-order Taylor series expansions of $T_a$, $T_b$, $T_c$, $T_d$:

$$\begin{aligned}
T_a(\boldsymbol{c}+\boldsymbol{h}) &= T_a(\boldsymbol{c}) + \langle h_a, \nabla_a T_a(\boldsymbol{c})\rangle_{\mathcal{H}} + \langle h_b, \nabla_b T_a(\boldsymbol{c})\rangle_{\mathcal{H}} + \ldots + \langle h_d, \nabla_d T_a(\boldsymbol{c})\rangle_{\mathcal{H}} \\
&\quad + o(\|\boldsymbol{h}\|_{\mathcal{H}^4}) \\
T_b(\boldsymbol{c}+\boldsymbol{h}) &= T_b(\boldsymbol{c}) + \langle h_a, \nabla_a T_b(\boldsymbol{c})\rangle_{\mathcal{H}} + \langle h_b, \nabla_b T_b(\boldsymbol{c})\rangle_{\mathcal{H}} + \ldots + \langle h_d, \nabla_d T_b(\boldsymbol{c})\rangle_{\mathcal{H}} \\
&\quad + o(\|\boldsymbol{h}\|_{\mathcal{H}^4}) \\
T_c(\boldsymbol{c}+\boldsymbol{h}) &= T_c(\boldsymbol{c}) + \langle h_a, \nabla_a T_c(\boldsymbol{c})\rangle_{\mathcal{H}} + \langle h_b, \nabla_b T_c(\boldsymbol{c})\rangle_{\mathcal{H}} + \ldots + \langle h_d, \nabla_d T_c(\boldsymbol{c})\rangle_{\mathcal{H}} \\
&\quad + o(\|\boldsymbol{h}\|_{\mathcal{H}^4}) \\
T_d(\boldsymbol{c}+\boldsymbol{h}) &= T_d(\boldsymbol{c}) + \langle h_a, \nabla_a T_d(\boldsymbol{c})\rangle_{\mathcal{H}} + \langle h_b, \nabla_b T_d(\boldsymbol{c})\rangle_{\mathcal{H}} + \ldots + \langle h_d, \nabla_d T_d(\boldsymbol{c})\rangle_{\mathcal{H}} \\
&\quad + o(\|\boldsymbol{h}\|_{\mathcal{H}^4})
\end{aligned}$$

where $\nabla_\iota$ is the gradient at $\boldsymbol{c}$ w.r.t. $f_\iota$ (Fréchet partial derivative).

Following the methodology shown by [5] in the complex case, and considering the quaternion properties outlined in [4], it can be shown that

$$\begin{aligned}
\boldsymbol{T}(\boldsymbol{c}+\boldsymbol{h}) &= \boldsymbol{T}(\boldsymbol{c}) + \left\langle \boldsymbol{h},\ \tfrac{1}{4}(\nabla_a \boldsymbol{T}(\boldsymbol{c}) - i\nabla_b \boldsymbol{T}(\boldsymbol{c}) - j\nabla_c \boldsymbol{T}(\boldsymbol{c}) - k\nabla_d \boldsymbol{T}(\boldsymbol{c}))^* \right\rangle_{\mathbb{Q}} \\
&\quad + \left\langle \boldsymbol{h}^i,\ \tfrac{1}{4}(\nabla_a \boldsymbol{T}(\boldsymbol{c}) - i\nabla_b \boldsymbol{T}(\boldsymbol{c}) + j\nabla_c \boldsymbol{T}(\boldsymbol{c}) + k\nabla_d \boldsymbol{T}(\boldsymbol{c}))^* \right\rangle_{\mathbb{Q}} \\
&\quad + \left\langle \boldsymbol{h}^j,\ \tfrac{1}{4}(\nabla_a \boldsymbol{T}(\boldsymbol{c}) + i\nabla_b \boldsymbol{T}(\boldsymbol{c}) - j\nabla_c \boldsymbol{T}(\boldsymbol{c}) + k\nabla_d \boldsymbol{T}(\boldsymbol{c}))^* \right\rangle_{\mathbb{Q}} \\
&\quad + \left\langle \boldsymbol{h}^k,\ \tfrac{1}{4}(\nabla_a \boldsymbol{T}(\boldsymbol{c}) + i\nabla_b \boldsymbol{T}(\boldsymbol{c}) + j\nabla_c \boldsymbol{T}(\boldsymbol{c}) - k\nabla_d \boldsymbol{T}(\boldsymbol{c}))^* \right\rangle_{\mathbb{Q}} + o(\|\boldsymbol{h}\|_{\mathbb{Q}})
\end{aligned} \tag{10.32}$$

From this equation, we can see the Taylor series expansion of $\boldsymbol{T}$ expressed in terms of $\boldsymbol{h}$, $\boldsymbol{h}^i$, $\boldsymbol{h}^j$ and $\boldsymbol{h}^k$. The same process may also be extended to higher order Taylor series.

Based on (10.32), the Fréchet-based $\mathbb{HR}$ gradients is

$$\begin{bmatrix} \nabla_f \boldsymbol{T}(f, f^i, f^j, f^k)(\boldsymbol{c}) \\ \nabla_{f^i} \boldsymbol{T}(f, f^i f^j f^k)(\boldsymbol{c}) \\ \nabla_{f^j} \boldsymbol{T}(f, f^i, f^j, f^k)(\boldsymbol{c}) \\ \nabla_{f^k} \boldsymbol{T}(f, f^i, f^j, f^k)(\boldsymbol{c}) \end{bmatrix} = \frac{1}{4} \begin{bmatrix} 1 & -i & -j & -k \\ 1 & -i & j & k \\ 1 & i & -j & k \\ 1 & i & j & -k \end{bmatrix} \begin{bmatrix} \nabla_{f_a} \boldsymbol{T}(f_a, f_b, f_c, f_d)(\boldsymbol{c}) \\ \nabla_{f_b} \boldsymbol{T}(f_a, f_b, f_c, f_d)(\boldsymbol{c}) \\ \nabla_{f_c} \boldsymbol{T}(f_a, f_b, f_c, f_d)(\boldsymbol{c}) \\ \nabla_{f_d} \boldsymbol{T}(f_a, f_b, f_c, f_d)(\boldsymbol{c}) \end{bmatrix} \tag{10.33}$$

and similarly, using the conjugate quaternion properties from Chap. 5, the Fréchet $\mathbb{HR}^*$ gradients may be shown to be

$$\begin{bmatrix} \nabla_{f^*} \boldsymbol{T}(f^*, f^{i*}, f^{j*}, f^{k*})(\boldsymbol{c}) \\ \nabla_{f^{i*}} \boldsymbol{T}(f^*, f^{i*}, f^{j*}, f^{k*})(\boldsymbol{c}) \\ \nabla_{f^{j*}} \boldsymbol{T}(f^*, f^{i*}, f^{j*}, f^{k*})(\boldsymbol{c}) \\ \nabla_{f^{k*}} \boldsymbol{T}(f^*, f^{i*}, f^{j*}, f^{k*})(\boldsymbol{c}) \end{bmatrix} = \frac{1}{4} \begin{bmatrix} 1 & i & j & k \\ 1 & i & -j & -k \\ 1 & -i & j & -k \\ 1 & -i & -j & k \end{bmatrix} \begin{bmatrix} \nabla_{f_a} \boldsymbol{T}(f_a, f_b f_c, f_d)(\boldsymbol{c}) \\ \nabla_{f_b} \boldsymbol{T}(f_a, f_b f_c, f_d)(\boldsymbol{c}) \\ \nabla_{\mathrm{f}_c} \boldsymbol{T}(f_a, f_b f_c, f_d)(\boldsymbol{c}) \\ \nabla_{\mathrm{f}_d} \boldsymbol{T}(f_a, f_b f_c, f_d)(\boldsymbol{c}) \end{bmatrix} \tag{10.34}$$

The Fréchet $\mathbb{HR}$ and $\mathbb{HR}^*$ gradients yielded properties:

1. If $\boldsymbol{T}(\boldsymbol{f}) = <\boldsymbol{h}, \boldsymbol{f}^* >_{\mathbb{Q}}$ or $\boldsymbol{T}(\boldsymbol{f}) = <\boldsymbol{f}, \boldsymbol{h}^* >_{\mathbb{Q}}$ then $\nabla_f \boldsymbol{T}(\boldsymbol{c}) = \boldsymbol{h}$ and $\nabla_{f^\eta} \boldsymbol{T}(\boldsymbol{c}) = 0$ for $\eta \in \{i, j, k\}$ at every $\boldsymbol{c}$. Also: $\nabla_{f^*} \boldsymbol{T}(\boldsymbol{c}) = -\boldsymbol{h}/2$ and $\nabla_{f^{\eta *}} \boldsymbol{T}(\boldsymbol{c}) = \boldsymbol{h}/2$ for $\eta \in \{i, j, k\}$.
2. If $\boldsymbol{T}(\boldsymbol{f}) = <\boldsymbol{h}, \boldsymbol{f} >_{\mathbb{Q}}$ or $\boldsymbol{T}(\boldsymbol{f}) = <\boldsymbol{f}^*, \boldsymbol{h}^* >_{\mathbb{Q}}$ then $\nabla_{f^*} \boldsymbol{T}(\boldsymbol{c}) = \boldsymbol{h}$ and $\nabla_{f^{\eta *}} \boldsymbol{T}(\boldsymbol{c}) = 0$ for $\eta \in \{i, j, k\}$ at every $\boldsymbol{c}$. Also: $\nabla_f \boldsymbol{T}(\boldsymbol{c}) = -\boldsymbol{h}/2$ and $\nabla_{f^\eta} \boldsymbol{T}(\boldsymbol{c}) = \boldsymbol{h}/2$ for $\eta \in \{i, j, k\}$.
3. If $\boldsymbol{T}(\boldsymbol{f}) = <\boldsymbol{h}, \boldsymbol{f}^i >_{\mathbb{Q}}$ or $\boldsymbol{T}(\boldsymbol{f}) = <\boldsymbol{f}^{i*}, \boldsymbol{h}^* >_{\mathbb{Q}}$ then $\nabla_{f^{i*}} \boldsymbol{T}(\boldsymbol{c}) = \boldsymbol{h}$ and $\nabla_{f^*} \boldsymbol{T}(\boldsymbol{c}) = \nabla_{f^{j*}} \boldsymbol{T}(\boldsymbol{c}) = \nabla_{f^{k*}} \boldsymbol{T}(\boldsymbol{c}) = 0$ at every $\boldsymbol{c}$. Also: $\nabla_{f^i} \boldsymbol{T}(\boldsymbol{c}) = -\boldsymbol{h}/2$ and $\nabla_{f^\eta} \boldsymbol{T}(\boldsymbol{c}) = \boldsymbol{h}/2$ for $\eta \in \{1, j, k\}$.
4. If $\boldsymbol{T}(\boldsymbol{f}) = <\boldsymbol{h}, \boldsymbol{f}^{i*} >_{\mathbb{Q}}$ or $\boldsymbol{T}(\boldsymbol{f}) = <\boldsymbol{f}^i, \boldsymbol{h}^* >_{\mathbb{Q}}$ then $\nabla_{f^i} \boldsymbol{T}(\boldsymbol{c}) = \boldsymbol{h}$ and $\nabla_f \boldsymbol{T}(\boldsymbol{c}) = \nabla_{f^j} \boldsymbol{T}(\boldsymbol{c}) = \nabla_{f^k} \boldsymbol{T}(\boldsymbol{c}) = 0$ at every $\boldsymbol{c}$. Also: $\nabla_{f^{i*}} \boldsymbol{T}(\boldsymbol{c}) = -\boldsymbol{h}/2$ and $\nabla_{f^{\eta *}} \boldsymbol{T}(\boldsymbol{c}) = \boldsymbol{h}/2$ for $\eta \in \{1, j, k\}$.
5. Product Rule: When the components of operators $\boldsymbol{T}, \boldsymbol{S} : \mathbb{Q} \to \mathbb{H}$ are all Fréchet differentiable in the real sense at $\boldsymbol{c} \in \mathbb{Q}$, then

$$\nabla_{f^*}(\boldsymbol{T} \cdot \boldsymbol{S})(\boldsymbol{c}) = [\nabla_{f^*} \boldsymbol{T}(\boldsymbol{c})] \boldsymbol{S}(\boldsymbol{c}) + \boldsymbol{T}(\boldsymbol{c}) [\nabla_{f^*} \boldsymbol{S}(\boldsymbol{c})]$$

(Note: Quaternion multiplication is non-commutative).

The properties described here apply to functions defined on Hilbert spaces, which differ from the $\mathbb{HR}$ calculus in [4].

Use of the Fréchet derivative allowed for providing the gradient definition on the desired quaternion Hilbert space, which is considered isomorphic with a multivariable real Hilbert space. For the multivariable real space, the partial Fréchet gradient with respect to each real component of the input is used to provide the Taylor series expansion of each of the real components of the nonlinear function output (i.e. equation before (10.32)). These partial gradients were used to express the Fréchet-based gradient with respect to a quaternion (i.e.$\mathbb{HR}$ calculus for Hilbert spaces) based on the isomorphism between $\mathbb{Q}$ and $\mathcal{H}^4$.

The resulting Fréchet $\mathbb{HR}$ and $\mathbb{HR}^*$ gradients naturally extend the $\mathbb{HR}$, $\mathbb{HR}^*$ gradients of [4] and their properties, although with inner product and transform space definitions on Hilbert spaces. The Fréchet-based gradients and properties are used to find the cost function gradients in the next section.

## 10.4 Derivation of Algorithms

To combine quaternions with kernel adaptive methods, we use an approach based on LMS, referred to here as the quaternion kernel LMS (or Quat-KLMS). The goal is to estimate the channel output where the reference data, i.e. desired response, is: $d(n) = f\ (\mathbf{x}(n)) + v(n)$. Here $f\,(\cdot)$ is an unknown nonlinear quaternion function.

The estimator form is

$$\hat{d}_{\mathrm{Q}}(n) = < \widehat{\boldsymbol{\Phi}}_{\mathrm{Q}}(\mathbf{x}(n)), \mathbf{w}_{\mathrm{Q}} >_{\mathrm{Q}} \tag{10.35}$$

where $\mathbf{x}(n)$ is the input vector and $\mathbf{w}_{\mathrm{Q}}$ is the weight vector, both with quaternion-valued components. The notations $\widehat{\boldsymbol{\Phi}}_{\mathrm{Q}}(\cdot)$ and $< \cdot, \cdot >_{\mathrm{Q}}$ are the same as described previously.

For combining widely linear estimation and Quat-KLMS, we consider here two methods. QWL estimation may be naturally applied after kernel mapping $\mathbf{x}(n)$ to $\widehat{\boldsymbol{\Phi}}_{Q}(\mathbf{x}(n))$, extending the linear estimation to a widely linear form (see (5.23), Chap. 5). We refer to this approach as the widely linear Quat-KLMS algorithm (WL-Quat-KLMS).

Alternatively, we consider an approach extending the input vector $\mathbf{x}(n)$ based on the perpendicular involutions shown in (10.23), i.e. forming $\mathbf{x}^{a}(n)$, before the kernel map. An interpretation of this approach is provided later. The approach we refer to as the involution-basis Quat-KLMS (IB-Quat-KLMS). The two methods are as shown.

**Method 1: WL-Quat-KLMS:**

Based on this approach, the estimation form is

$$\hat{d}_{\mathrm{Q}_{WL}}(n) = < \widehat{\boldsymbol{\Phi}}_{\mathrm{Q}_{WL}}(\mathbf{x}(n)), \mathbf{w}_{\mathrm{Q}_{WL}} >_{\mathrm{Q}_{WL}} \tag{10.36}$$

where the mapping $\widehat{\boldsymbol{\Phi}}_{\mathrm{Q}_{WL}}(\cdot)$ may be defined in terms of mapping $\widehat{\boldsymbol{\Phi}}_{\mathrm{Q}}(\cdot)$ as $\widehat{\boldsymbol{\Phi}}_{\mathrm{Q}_{WL}}(\mathbf{a}) = \left[\widehat{\Phi}_{\mathrm{Q}}^{T}(\mathbf{a})\; \widehat{\boldsymbol{\Phi}}_{\mathrm{Q}}^{iI}(\mathbf{a})\; \widehat{\boldsymbol{\Phi}}_{\mathrm{Q}}^{jT}(\mathbf{a})\; \widehat{\boldsymbol{\Phi}}_{\mathrm{Q}}^{kT}(\mathbf{a})\right]^{T}$, and $\mathbf{w}_{\mathrm{Q}_{WL}}$ is a weight vector with quaternion-valued components. $\mathrm{Q}_{WL}$ denotes the associated quaternion RKHS with the mapping $\widehat{\boldsymbol{\Phi}}_{\mathrm{Q}_{WL}}(\cdot)$.

**Method 2: IB-Quat-KLMS:**

Based on this approach, we may write

$$\hat{d}_{\mathrm{Q}_{IB}}(n) = < \widehat{\boldsymbol{\Phi}}_{\mathrm{Q}_{IB}}(\mathbf{x}(n)), \mathbf{w}_{\mathrm{Q}_{IB}} >_{\mathrm{Q}_{IB}} \tag{10.37}$$

where mapping $\widehat{\boldsymbol{\Phi}}_{\mathrm{Q}_{IB}}(\cdot)$ may be defined based on $\widehat{\boldsymbol{\Phi}}_{\mathrm{Q}}(\cdot)$ as $\widehat{\boldsymbol{\Phi}}_{\mathrm{Q}_{IB}}(\mathbf{a}) = \widehat{\boldsymbol{\Phi}}_{\mathrm{Q}}\left(\left[\mathbf{a}^{T}\; \mathbf{a}^{iT}\; \mathbf{a}^{jT}\; \mathbf{a}^{kT}\right]^{T}\right)$. Note that the dimensionality of mapping $\widehat{\boldsymbol{\Phi}}_{\mathrm{Q}}(\cdot)$ may differ depending on the input vector length, $d$ (see (10.3), (10.4)). The weight vector $\mathbf{w}_{\mathrm{Q}_{IB}}$ is of suitable dimension size to perform the inner product. $\mathrm{Q}_{IB}$ denotes the quaternion RKHS associated with $\widehat{\boldsymbol{\Phi}}_{\mathrm{Q}_{IB}}(\cdot)$.

We now consider the weight updates of these algorithms (referred to as Quat-KLMS, WL-Quat-KLMS and IB-Quat-KLMS here), as well as their final algorithm forms suitable for implementation.

### 10.4.1 Quat-KLMS Algorithm

For the Quat-KLMS, the goal of the algorithm is finding weights, based on the estimator (10.35), minimizing the mean square error, $E[|e(n)|^2]$, where $e(n) = d(n) - \hat{d}_Q(n)$. With the LMS-based approach, the instantaneous square error $|e(n)|^2$ is used instead of the expected value of the cost function.

A normalized form for the Quat-KLMS weight update is

$$\mathbf{w}_Q(n) = \mathbf{w}_Q(n-1) - \frac{\eta \nabla_{\mathbf{w}_Q^*}\{|e(n)|^2\}}{\left\langle \widehat{\Phi}_Q(\mathbf{x}(n)), \widehat{\mathbf{\Phi}}_Q(\mathbf{x}(n)) \right\rangle_Q} \tag{10.38}$$

where $\nabla_{\mathbf{w}_Q^*}$ is the gradient with respect to $\mathbf{w}_Q{}^*$, and $\eta$ is step size. Note that this quaternion-based gradient should be found with respect to the conjugate of the weights following Branchwood's gradient (see [4]), similar to the complex case.

To obtain the update, we evaluate $\nabla_{\mathbf{w}_{Q}*}\{|e(n)|^2\}$, yielding

$$\begin{aligned}
\nabla_{\mathbf{w}_{Q}*}\{|e(n)|^2\} &= \nabla_{\mathbf{w}_Q}\left\{\left|d(n) - < \widehat{\mathbf{\Phi}}(\mathbf{x}(n)), \mathbf{w}_Q >_Q\right|^2\right\} \\
&= \nabla_{\mathbf{w}_Q}\left\{\left(d(n) - < \widehat{\mathbf{\Phi}}_Q(\mathbf{x}(n)), \mathbf{w}_Q >_Q\right)\right. \\
&\quad \left.\left(d(n) - < \widehat{\mathbf{\Phi}}_Q(\mathbf{x}(n)), \mathbf{w}_Q >_Q\right)^*\right\} \\
&= \nabla_{\mathbf{w}_Q} * \left\{\left(d(n) - < \widehat{\mathbf{\Phi}}_Q(\mathbf{x}(n)), \mathbf{w}_Q >_Q\right)\right. \\
&\quad \left.\left(d^*(n) - < \mathbf{w}_Q, \widehat{\mathbf{\Phi}}_Q(\mathbf{x}(n)) >_Q\right)\right\}
\end{aligned} \tag{10.39}$$

Consider that the gradient in the case here is performed on a real-valued cost function defined on a quaternion Hilbert space (specifically, an RKHS). Here we require the modified $\mathbb{HR}$ calculus and properties from Sect. 10.3.

Based on these gradient properties, we get

$$\nabla_{\mathbf{w}_Q;}\left\{|e(n)|^2\right\} = -\widehat{\mathbf{\Phi}}_Q(\mathbf{x}(n))e^*(n) + \frac{1}{2}e(n)\widehat{\mathbf{\Phi}}_Q^*(\mathbf{x}(n)) \tag{10.40}$$

and the weight update for Quat-KLMS may be expressed as

$$\mathbf{w}_Q(n) = \mathbf{w}_Q(n-1) + \eta\left(\widehat{\Phi}_Q(\mathbf{x}(n))e^*(n) - \frac{1}{2}e(n)\widehat{\mathbf{\Phi}}_Q^*(\mathbf{x}(n))\right)/\mathrm{K}_Q(n)$$

where $\mathrm{K}_S(a) = \left\langle \widehat{\Phi}_S(\mathbf{x}(a)), \widehat{\mathbf{\Phi}}_S(\mathbf{x}(a)) \right\rangle_S$ is the normalization term.

Assuming $\mathbf{w}(0) = 0$, the weights at iteration $n$ can be written as

$$\mathbf{w}_Q(n) = \eta \sum_{m=1}^{n} \left[ \left( \widehat{\Phi}_Q(\mathbf{x}(m))e^*(m) - \frac{1}{2}e(m)\widehat{\mathbf{\Phi}}_Q^*(\mathbf{x}(m)) \right) \mathrm{K}_Q^{-1}(m) \right] \tag{10.41}$$

The estimates at iteration $n$ would then be

$$\begin{aligned}
\hat{d}_Q(n) &= < \widehat{\mathbf{\Phi}}_Q(\mathbf{x}(n)), \mathbf{w}_Q(n-1) >_Q \\
&= \left[ \eta \sum_{m=1}^{n-1} \left( \widehat{\Phi}_Q(\mathbf{x}(m))e^*(m) - \frac{1}{2}e(m)\widehat{\mathbf{\Phi}}_Q^*(\mathbf{x}(m)) \right) \mathrm{K}_Q^{-1}(m) \right]^H \widehat{\mathbf{\Phi}}_Q(\mathbf{x}(n)) \\
&= \left[ \eta \sum_{m=1}^{n-1} \mathrm{K}_Q^{-1}(m) \left( e(m)\widehat{\mathbf{\Phi}}_Q^H(\mathbf{x}(m)) - \frac{1}{2}\widehat{\mathbf{\Phi}}_Q^T(\mathbf{x}(m))e^*(m) \right) \widehat{\mathbf{\Phi}}_Q(\mathbf{x}(n)) \right]
\end{aligned}$$

where, in the last step, ${\mathrm{K}_Q}^{-1}(m)$ is Hermitian.

We observe that, due to the non-commutative property of quaternion multiplication, the algorithm may not easily be cast in inner products using the well-known "kernel trick" [5, 6]. To form the gradient so that we may use inner products, we rewrite the gradient as follows. We equivalently write (10.40) as

$$\nabla^*_{\mathbf{w}_e}\left\{|e(n)|^2\right\} = -\frac{3}{2}\widehat{\mathbf{\Phi}}_Q(\mathbf{x}(n))e^*(n) + \frac{1}{2}\left[\widehat{\Phi}_Q(\mathbf{x}(n))e^*(n) + e(n)\widehat{\mathbf{\Phi}}_Q^*(\mathbf{x}(n))\right] \tag{10.42}$$

Since $a + a^* = 2\mathrm{Re}\{a\}$, we can write the second term of (10.42) as

$$\begin{aligned}
\mathrm{Re}\left\{\widehat{\Phi}_Q(\mathbf{x}(n))e^*(n)\right\} &= \widehat{\mathbf{\Phi}}_a(\mathbf{x}(n))e_a(n) + \widehat{\mathbf{\Phi}}_b(\mathbf{x}(n))e_b(n) \\
&\quad + \widehat{\mathbf{\Phi}}_c(\mathbf{x}(n))e_c(n) + \widehat{\mathbf{\Phi}}_d(\mathbf{x}(n))e_d(n)
\end{aligned}$$

where $\widehat{\mathbf{\Phi}}_a(\cdot)$ to $\widehat{\mathbf{\Phi}}_d(\cdot)$ and $e_a(n)$ to $e_d(n)$ are the components of $\widehat{\mathbf{\Phi}}_Q(\cdot)$ and $e(n)$, respectively.

From the quaternion-based properties of Chap. 5, we get

$$\begin{aligned}
\mathrm{Re}\left\{\widehat{\Phi}_Q(\mathbf{x}(n))e^*(n)\right\} &= \frac{1}{4}\left[\widehat{\Phi}_Q(\mathbf{x}(n)) + \widehat{\mathbf{\Phi}}_Q^i(\mathbf{x}(n)) + \widehat{\mathbf{\Phi}}_Q^j(\mathbf{x}(n)) + \widehat{\mathbf{\Phi}}_Q^k(\mathbf{x}(n))\right]e_a(n) \\
&\quad - \frac{1}{4}\left[\widehat{\Phi}_Q(\mathbf{x}(n)) + \widehat{\mathbf{\Phi}}_Q^i(\mathbf{x}(n)) - \widehat{\mathbf{\Phi}}_Q^j(\mathbf{x}(n)) - \widehat{\mathbf{\Phi}}_Q^k(\mathbf{x}(n))\right]ie_b(n) \\
&\quad - \frac{1}{4}\left[\widehat{\Phi}_Q(\mathbf{x}(n)) - \widehat{\mathbf{\Phi}}_Q^i(\mathbf{x}(n)) + \widehat{\mathbf{\Phi}}_Q^j(\mathbf{x}(n)) - \widehat{\mathbf{\Phi}}_Q^k(\mathbf{x}(n))\right]je_c(n) \\
&\quad - \frac{1}{4}\left[\widehat{\Phi}_Q(\mathbf{x}(n)) - \widehat{\mathbf{\Phi}}_Q^i(\mathbf{x}(n)) - \widehat{\mathbf{\Phi}}_Q^j(\mathbf{x}(n)) + \widehat{\mathbf{\Phi}}_Q^k(\mathbf{x}(n))\right]ke_d(n)
\end{aligned} \tag{10.43}$$

Finally, using the conjugate quaternion-based properties to factor yields

$$\mathrm{Re}\left\{\widehat{\Phi}_{\mathrm{Q}}(\mathbf{x}(n))e^{*}(n)\right\} =$$
$$\frac{1}{4}\left[\widehat{\Phi}_{\mathrm{Q}}(\mathbf{x}(n))e^{*}(n) + \widehat{\mathbf{\Phi}}^{i}_{\mathrm{Q}}(\mathbf{x}(n))e^{i^{*}}(n) + \widehat{\mathbf{\Phi}}^{j}_{\mathrm{Q}}(\mathbf{x}(n))e^{j^{*}}(n) + \widehat{\mathbf{\Phi}}^{k}_{\mathrm{Q}}(\mathbf{x}(n))e^{k^{*}}(n)\right] \tag{10.44}$$

and the gradient becomes

$$\begin{aligned}\nabla_{\mathbf{w}_{e*}}\{|e(n)|^2\} &= -\frac{5}{4}\widehat{\mathbf{\Phi}}_{\mathrm{Q}}(\mathbf{x}(n))e^{*}(n) + \frac{1}{4}\widehat{\mathbf{\Phi}}^{i}_{\mathrm{Q}}(\mathbf{x}(n))e^{i^{*}}(n)\\ &+ \frac{1}{4}\widehat{\mathbf{\Phi}}^{j}_{\mathrm{Q}}(\mathbf{x}(n))e^{j^{*}}(n) + \frac{1}{4}\widehat{\mathbf{\Phi}}^{k}_{\mathrm{Q}}(\mathbf{x}(n))e^{k^{*}}(n)\end{aligned} \tag{10.45}$$

Therefore, (10.41) can be equivalently expressed as

$$\begin{aligned}\mathbf{w}_{\mathrm{Q}}(n) = \eta\sum_{m=1}^{n}\Bigg[&\frac{5}{4}\widehat{\mathbf{\Phi}}_{\mathrm{Q}}(\mathbf{x}(m))e^{*}(m) - \frac{1}{4}\widehat{\mathbf{\Phi}}_{\mathrm{Q}}\big(\mathbf{x}^{i}(m)\big)e^{i^{*}}(m)\\ &-\frac{1}{4}\widehat{\mathbf{\Phi}}_{\mathrm{Q}}\big(\mathbf{x}^{j}(m)\big)e^{j^{*}}(m) - \frac{1}{4}\widehat{\mathbf{\Phi}}_{\mathrm{Q}}\big(\mathbf{x}^{k}(m)\big)e^{k^{*}}(m)\Bigg]\mathrm{K}_{\mathrm{Q}}^{-1}(m)\end{aligned}$$

where we use assumption: $\widehat{\mathbf{\Phi}}^{\eta}_{\mathrm{Q}}(\mathbf{x}(n)) = \widehat{\mathbf{\Phi}}_{\mathrm{Q}}(\mathbf{x}^{\eta}(n))$ for $\eta \in \{i, k, j\}$ as described in Sect. 10.2.1.

Based on this, the estimates in this case are

$$\begin{aligned}\hat{d}_{\mathrm{Q}}(n) &=< \widehat{\mathbf{\Phi}}_{\mathrm{Q}}(\mathbf{x}(n)), \mathbf{w}_{\mathrm{Q}}(n-1) >_{\mathrm{Q}}\\ &= \eta\sum_{m=1}^{n-1}\mathrm{K}_{\mathrm{Q}}^{-1}(m)\Bigg[\frac{5}{4}e(m)\kappa_{\mathrm{Q}}(\mathbf{x}(m), \mathbf{x}(n)) - \frac{1}{4}e^{i}(m)\kappa_{\mathrm{Q}}\big(\mathbf{x}^{i}(m), \mathbf{x}(n)\big)\\ &-\frac{1}{4}e^{j}(m)\kappa_{\mathrm{Q}}\big(\mathbf{x}^{j}(m), \mathbf{x}(n)\big) - \frac{1}{4}e^{k}(m)\kappa_{\mathrm{Q}}\big(\mathbf{x}^{k}(m), \mathbf{x}(n)\big)\Bigg]\end{aligned} \tag{10.46}$$

where $\kappa_{\mathrm{Q}}(\cdot, \cdot)$ is the quaternion Gaussian kernel.

For the quaternion-extended Gaussian kernel, we see, considering the RKHS from Sect. 10.2.2:

$$\widehat{\mathbf{\Phi}}^{i}_{\mathrm{Q}}(\mathbf{a}) = \widehat{\mathbf{\Phi}}_{\mathrm{Q}}(\mathbf{a})(-j) \quad \widehat{\mathbf{\Phi}}^{j}_{\mathrm{Q}}(\mathbf{a}) = \widehat{\mathbf{\Phi}}_{\mathrm{Q}}(\mathbf{a})(-k) \quad \widehat{\mathbf{\Phi}}^{k}_{\mathrm{Q}}(\mathbf{a}) = \widehat{\mathbf{\Phi}}_{\mathrm{Q}}(\mathbf{a})(-i)$$

and the Quat-KLMS may be expressed as

$$\hat{d}_{\mathrm{Q}}(n) =< \widehat{\mathbf{\Phi}}_{\mathrm{Q}}(\mathbf{x}(n)), \mathbf{w}_{\mathrm{Q}}(n-1) >_{\mathrm{Q}}$$

$$
\begin{aligned}
&= \eta \sum_{m=1}^{n-1} \mathrm{K}_{\mathrm{Q}}^{-1}(m) \left[ \frac{5}{4} e(m) \kappa_{\mathrm{Q}}(\mathbf{x}(m), \mathbf{x}(n)) - \frac{1}{4} e^{i}(m) j \kappa_{\mathrm{Q}}(\mathbf{x}(m), \mathbf{x}(n)) \right. \\
&\quad \left. - \frac{1}{4} e^{j}(m) k \kappa_{\mathrm{Q}}(\mathbf{x}(m), \mathbf{x}(n)) - \frac{1}{4} e^{k}(m) i \kappa_{\mathrm{Q}}(\mathbf{x}(m), \mathbf{x}(n)) \right] \\
&= \frac{\eta}{4} \sum_{m=1}^{n-1} \mathrm{K}_{\mathrm{Q}}^{-1}(m) \left[ \left( 5e(m) - e^{i}(m) j - e^{j}(m) k - e^{k}(m) i \right) \kappa_{\mathrm{Q}}(\mathbf{x}(m), \mathbf{x}(n)) \right]
\end{aligned}
\tag{10.47}
$$

where $\kappa_{\mathrm{Q}}(\cdot, \cdot)$ is the Quat-Ext Gaussian kernel.

An alternative approach for deriving the Quat-Ext kernel-based algorithm can be obtained by viewing the RKHS in terms of a real-valued mapping $\Phi(\mathbf{x}_{comp}(n))$ and quaternion scalar $s_{\Phi}$ (both defined in Sect. 10.2.2). This approach yields

$$
\begin{aligned}
\hat{d}_{\mathrm{Q}}(n) &= \eta \left[ \sum_{m=1}^{n} \left( e(m) \left( s_{\Phi} \Phi(\mathbf{x}_{\mathrm{comp}}(m)) \right)^{H} - \frac{1}{2} \left( s_{\Phi} \Phi(\mathbf{x}_{\mathrm{comp}}(m)) \right)^{T} e^{*}(m) \right) \right] \\
&\quad s_{\Phi} \Phi(\mathbf{x}_{\mathrm{comp}}(n)) \\
&= \eta \left[ \sum_{m=1}^{n} \left( e(m) \Phi^{T}(\mathbf{x}_{\mathrm{comp}}(m)) s_{\Phi}^{*} - \frac{1}{2} \Phi^{T}(\mathbf{x}_{\mathrm{comp}}(m)) s_{\Phi} e^{*}(m) \right) \right] \\
&\quad s_{\Phi} \Phi(\mathbf{x}_{\mathrm{comp}}(n)) \\
&= \eta \left[ \sum_{m=1}^{n} \left( e(m) s_{\Phi}^{*} - \frac{1}{2} s_{\Phi} e^{*}(m) \right) \Phi^{T}(\mathbf{x}_{\mathrm{comp}}(m)) \right] s_{\Phi} \Phi(\mathbf{x}_{\mathrm{comp}}(n)) \\
&= \eta \sum_{m=1}^{n} \left[ \left( e(m) |s_{\Phi}|^{2} - \frac{1}{2} s_{\Phi} e^{*}(m) s_{\Phi} \right) \Phi^{T}(\mathbf{x}_{\mathrm{comp}}(m)) \Phi(\mathbf{x}_{\mathrm{comp}}(n)) \right] \\
&= \eta \sum_{m=1}^{n} \left[ \left( e(m) |s_{\Phi}|^{2} - \frac{1}{2} s_{\Phi} e^{*}(m) s_{\Phi} \right) \kappa_{\sigma, \mathbb{R}^{d}}(\mathbf{x}_{\mathrm{comp}}(n), \mathbf{x}_{\mathrm{comp}}(m)) \right]
\end{aligned}
\tag{10.48}
$$

where $\kappa_{\sigma,\mathbb{R}^d}(\cdot, \cdot)$ is the real Gaussian kernel form. Note normalization is not required, since $\mathrm{K}_{\mathrm{Q}}(a) = 4$ for all $a$ with the Quat-Ext Gaussian kernel. However, our analysis here focuses on (10.46) and (10.47), which form the Quat-KLMS estimates with the quaternion Gaussian and Quat-Ext Gaussian kernels.

We now describe the widely linear based algorithms. For these methods, we only consider the use of the quaternion Gaussian kernel. The reasoning is shown later in this section.

### 10.4.2 WL-Quat-KLMS Algorithm

For WL-Quat-KLMS, we minimize the mean square error $E[|e(n)|^2]$, where $e(n) = d(n) - \hat{d}_{Q_{WL}}(n)$, and $\hat{d}_{Q_{WL}}(n)$ is from (10.36).

Using a normalized LMS-based approach:

$$\mathbf{w}_{Q_{WL}}(n) = \mathbf{w}_{Q_{WL}}(n-1) - \frac{\eta \nabla_{\mathbf{w}_{Q_{WL}}*}\left\{|e(n)|^2\right\}}{K_{Q_{WL}}(n)} \tag{10.49}$$

where $\nabla_{\mathbf{w}_{QWL}*}$ is the gradient with respect to $\mathbf{w}_{QWL}{}^*$. This yields

$$\begin{aligned}\nabla_{\mathbf{w}_{\underline{Q}_{WL}}*}\left\{|e(n)|^2\right\} &= \nabla_{\mathbf{w}_{\underline{Q}_{WL}}*}\left\{\left|d(n) - <\widehat{\mathbf{\Phi}}_{Q_{WL}}(\mathbf{x}(n)), \mathbf{w}_{Q_{WL}} > Q_{Q_{WL}}\right|^2\right\} \\ &= \nabla_{\mathbf{w}_{Q_{WL}}*}\left\{\left(d(n) - <\widehat{\mathbf{\Phi}}_{Q_{WL}}(\mathbf{x}(n)), \mathbf{w}_{Q_{WL}} > Q_{Q_{WL}}\right)\right. \\ &\left.\times\left(d^*(n) - <\mathbf{w}_{Q_{WL}}, \widehat{\mathbf{\Phi}}_{Q_{WL}}(\mathbf{x}(n)) >_{Q_{WL}}\right)\right\}\end{aligned}$$

It can be easily seen that this gradient is

$$\nabla_{\mathbf{w}_{Q_{WL}}*}\left\{|e(n)|^2\right\} = -\widehat{\mathbf{\Phi}}_{Q_{WL}}(\mathbf{x}(n))e^*(n) + \frac{1}{2}e(n)\widehat{\mathbf{\Phi}}^*_{Q_{WL}}(\mathbf{x}(n)) \tag{10.50}$$

and the estimates, similar to before, are

$$\begin{aligned}\hat{d}_{Q_{WL}}(n) &= <\widehat{\mathbf{\Phi}}_{Q_{WL}}(\mathbf{x}(n)), \mathbf{W}_{Q_{WL}}(n-1) >_{Q_{WL}} \\ &= \eta \sum_{m=1}^{n-1} \mathrm{K}^{-1}_{Q_{WL}}(m)\left[\frac{5}{4}e(m)\kappa_{Q_{WL}}(\mathbf{x}(m), \mathbf{x}(n)) - \frac{1}{4}e^i(m)\kappa_{Q_{WL}}\left(\mathbf{x}^i(m), \mathbf{x}(n)\right)\right. \\ &\left.-\frac{1}{4}e^j(m)\kappa_{Q_{WL}}\left(\mathbf{x}^j(m), \mathbf{x}(n)\right) - \frac{1}{4}e^k(m)\kappa_{Q_{WL}}\left(\mathbf{x}^k(m), \mathbf{x}(n)\right)\right]\end{aligned} \tag{10.51}$$

where again we assume: $\widehat{\mathbf{\Phi}}^{\eta}_{Q}(\mathbf{x}(n)) = \widehat{\mathbf{\Phi}}_Q(\mathbf{x}^{\eta}(n))$ for $\eta \in \{i, k, j\}$. Function $\kappa_{Q_{WL}}(\mathbf{a}, \mathbf{b})$ can be evaluated using

$$\begin{aligned}\kappa_{Q_{WL}}(\mathbf{a}, \mathbf{b}) &= <\widehat{\mathbf{\Phi}}_{Q_{WL}}(\mathbf{b}), \widehat{\mathbf{\Phi}}_{Q_{WL}}(\mathbf{a}) >_{Q_{WL}} \\ &= \kappa_Q(\mathbf{a}, \mathbf{b}) + \kappa^i_Q(\mathbf{a}, \mathbf{b}) + \kappa^j_Q(\mathbf{a}, \mathbf{b}) + \kappa^k_Q(\mathbf{a}, \mathbf{b})\end{aligned}$$

where $a^{\eta}b^{\eta} = (ab)^{\eta}$ for $\eta \in \{i, k, j\}$ was used in factoring $\kappa_{Q_{WL}}(\mathbf{a}, \mathbf{b})$.

### 10.4.3 IB-Quat-KLMS Algorithm

For IB-Quat-KLMS, $E[|e(n)|^2]$ is minimized based on the estimator $\hat{d}_{Q_{IB}}(n)$ of (10.37). Using the same process shown for Quat-KLMS and WL-Quat-KLMS, the estimates are

$$\begin{aligned}\hat{d}_{Q_{IB}}(n) &= < \hat{\Phi}_{Q_{IB}}(\mathbf{x}(n)), \mathbf{w}_{Q_{IB}} >_{Q_{IB}} \\ &= \eta \sum_{m=1}^{n} \mathrm{K}_{Q_{IB}}^{-1}(n) \left[ \frac{5}{4} e(m) \kappa_{Q_{IB}}(\mathbf{x}(m), \mathbf{x}(n)) - \frac{1}{4} e^{i}(m) \kappa_{Q_{IB}}(\mathbf{x}^{i}(m), \mathbf{x}(n)) \right. \\ &\left. - \frac{1}{4} e^{j}(m) \kappa_{Q_{IB}}(\mathbf{x}^{j}(m), \mathbf{x}(n)) - \frac{1}{4} e^{k}(m) \kappa_{Q_{IB}}(\mathbf{x}^{k}(m), \mathbf{x}(n)) \right] \end{aligned} \tag{10.52}$$

where

$$\kappa_{Q_{IB}}(\mathbf{a}, \mathbf{b}) = \kappa_{Q}\left( \left[ \mathbf{a}^{T}\ \mathbf{a}^{iT}\ \mathbf{a}^{jT}\ \mathbf{a}^{kT} \right]^{T} \left[ \mathbf{b}^{T}\ \mathbf{b}^{iT}\ \mathbf{b}^{jT}\ \mathbf{b}^{kT} \right]^{T} \right)$$

From Eqs. (10.46), (10.51) and (10.52), the difference between Quat-KLMS, WL-Quat-KLMS, and IB-Quat-KLMS are the functions $\kappa_{Q}(\mathbf{a}, \mathbf{b})$, $\kappa_{Q_{WL}}(\mathbf{a}, \mathbf{b})$ and $\kappa_{Q_{IB}}(\mathbf{a}, \mathbf{b})$. These functions correspond to different transform spaces, based on their associated mappings $\widehat{\mathbf{\Phi}}_{Q}(\cdot)$, $\widehat{\mathbf{\Phi}}_{Q_{WL}}(\cdot)$ and $\widehat{\mathbf{\Phi}}_{Q_{IB}}(\cdot)$. In particular $\widehat{\mathbf{\Phi}}_{Q_{WL}}(\cdot)$ and $\widehat{\mathbf{\Phi}}_{Q_{IB}}(\cdot)$ are extended transform spaces based on widely linear estimation. A brief convergence analysis of the various algorithms described here is listed in Appendix F.

To understand the benefit of these transform spaces in learning nonlinear quaternion functions, we provide here interpretations of the various estimation approaches described.

### 10.4.4 Benefits of WL-Based Methods

From before, the mapping $\widehat{\mathbf{\Phi}}_{Q_{WL}}(\cdot)$ is defined as

$$\widehat{\mathbf{\Phi}}_{Q_{WL}}(\mathbf{a}) = \left[ \widehat{\Phi}_{Q}^{T}(\mathbf{a})\ \widehat{\Phi}_{Q}^{jT}(\mathbf{a})\ \widehat{\Phi}_{Q}^{jT}(\mathbf{a})\ \widehat{\Phi}_{Q}^{kT}(\mathbf{a}) \right]^{T}$$

We can see from the mapping that widely linear estimation with WL-Quat-KLMS is applied after the nonlinear transform, and resembles estimator (10.23) in the linear stage.

Meanwhile, mapping $\widehat{\mathbf{\Phi}}_{Q_{IB}}(\cdot)$ of IB-Quat-KLMS is defined as

$$\widehat{\mathbf{\Phi}}_{Q_{IB}}(\mathbf{a}) = \widehat{\mathbf{\Phi}}_{Q}\left( \left[ \mathbf{a}^{T}\ \mathbf{a}^{iT}\ \mathbf{a}^{jT}\ \mathbf{a}^{kT} \right]^{T} \right)$$

For this mapping, the input vector is formed from the original input concatenated with involutions. The estimator is

$$\begin{aligned}\hat{d}_{Q_{IB}}(n) &=< \widehat{\boldsymbol{\Phi}}_{Q_{IB}}(\mathbf{x}(n)), \mathbf{w}_{Q_{IB}} >_{Q_{IB}} \\ &=< \widehat{\boldsymbol{\Phi}}_{Q}\left(\left[\mathbf{x}^T(n)\,\mathbf{x}^{iT}(n)\,\mathbf{x}^{jT}(n)\,\mathbf{x}^{kT}(n)\right]^T\right)\mathbf{w}_{Q_{IB}} >_{Q}\end{aligned}$$

A possible interpretation of these mappings is as follows. Initially, consider MMSE (minimum mean-square error) estimation with real-valued data:

$$\hat{d} = E[d|\mathbf{x}]$$

with desired response $d = \mathbf{w}^H\mathbf{x} + v$, where $\mathbf{w}$ is the weight vector, and $v$ is the added noise.

For quaternions, when linear regression is considered based on the real-valued components, we may write

$$\hat{d}_{\{m\}} = E[d_{\{m\}}|\mathbf{x}_a,\ \mathbf{x}_b,\ \mathbf{x}_c,\ \mathbf{x}_d]$$

with $m \in \{a, b, c, d\}$, $d \in \mathbb{H}$ and $\mathbf{x} \in \mathbb{H}^r$, where $r$ is the length of observation vector $\mathbf{x}$. This can be equivalently expressed with the perpendicular involutions as a basis [4]

$$\hat{d}_{\{m\}} = E[d_{\{m\}}|\mathbf{x},\ \mathbf{x}^i,\ \mathbf{x}^j,\ \mathbf{x}^k] \tag{10.53}$$

From [11], although the estimate $\hat{d} = E[d|\mathbf{x}]$ is valid, the regression is linear when $d$, $\mathbf{x}$ are jointly normal and real. When complex normal data is used (or quaternion normal here), the regression required is widely linear. Equation (10.53) illustrates this in a similar manner with [4].

For the nonlinear case, the desired response is $d = f(\mathbf{x}) + v$, where $f(\cdot)$ is a nonlinear function. This may also be expressed $d = \mathbf{w}^H\widehat{\boldsymbol{\Phi}}_Q(x) + v$, where $\widehat{\boldsymbol{\Phi}}_Q(\cdot)$ is the transform map of a suitable quaternion kernel function.

It follows that the estimation may be described

$$\hat{d} = E\left[d\middle|\widehat{\Phi}_Q(\mathbf{x}),\ \widehat{\boldsymbol{\Phi}}^i_Q(\mathbf{x}),\ \widehat{\boldsymbol{\Phi}}^j_Q(\mathbf{x}),\ \widehat{\boldsymbol{\Phi}}^k_Q(\mathbf{x})\right] \tag{10.54}$$

illustrating widely linear regression. The WL-Quat-KLMS follows the required regression.

We note, however, that the use of the quaternion-extended (Quat-Ext) Gaussian kernel form may be viewed as

$$\hat{d} = E[d|\Phi(\mathbf{x}_a,\ \mathbf{x}_b,\ \mathbf{x}_c,\ \mathbf{x}_d)] \tag{10.55}$$

considering how the kernel function is formed in Sect. 10.2.2. The approach is based on using a real mapping $\Phi(\cdot)$ to learn the nonlinear function $f(\cdot)$ applied to components of quaternion $\mathbf{x}$.

Based on using perpendicular involutions as an equivalent basis, we consider

$$\hat{d} = E\Big[d\Big|\bar{\Phi}_{\mathrm{Q}}\big(\mathbf{x}, \mathbf{x}^{i}, \mathbf{x}^{j}, \mathbf{x}^{k}\big)\Big] \tag{10.56}$$

Equation (10.56) may be seen to describe the IB-Quat-KLMS.

The use of the perpendicular involutions of $\mathbf{x}$ to form the input to the nonlinear stage of the quaternion Gaussian kernel may be considered to yield a similar benefit observed with the Quat-Ext Gaussian kernel. Specifically, here, involutions are used instead of components. From [12], the complexified kernel form may be viewed as a dual real channel (DRC) formulation where a real kernel is applied to each channel. This form was seen in [12] to only differ from its augmented, or widely linear, extension by a scaling factor. We assume here that the Quat-Ext kernel provides a similar benefit.

A complete study of the benefit of the approach of (10.56) in forming an extended RKHS for quaternion-valued data (versus an extended RKHS based on the real components).

### 10.4.5 iQuat-KLMS Algorithm

An alternative Quat-KLMS algorithm can also be formed from the i-gradient (see Chap. 5 for details). For evaluating (10.39), considering that

$$\frac{\partial e(n)}{\partial \mathbf{w}_{Q}^{\beta}} = -\frac{1}{2}\widehat{\boldsymbol{\Phi}}(\mathbf{x}(n)) \quad \frac{\partial e^{*}(n)}{\partial \mathbf{w}_{Q}^{\beta}} = 0 \quad \forall\beta \in \{i, j, k\}$$

Using the gradient form (10.18), as well as Property 5 of Sect. 10.3, we obtain

$$\nabla_{\mathbf{w}_{Q}*}\{|e(n)|^2\} \;=\; -\tfrac{3}{2}\widehat{\boldsymbol{\Phi}}(\mathbf{x}(n))e^{*}(n)$$

The weight update becomes

$$\mathbf{w}_{Q}(n) = \mathbf{w}_{Q}(n-1) + \eta\Big(\tfrac{3}{2}\widehat{\boldsymbol{\Phi}}(\mathbf{x}(n))e^{*}(n)\Big)$$

and for iteration $n$ (assuming $\mathbf{w}(0) = 0$) we get

$$\mathbf{w}_{Q}(n) = \eta\sum_{m=1}^{n}\Big(\widehat{\Phi}(\mathbf{x}(n))e^{*}(n)\Big)$$

where the 3/2 scaling is included in step-size $\eta$.

The i-gradient-based estimates become

$$\hat{d}_Q(n) =< \widehat{\mathbf{\Phi}}(\mathbf{x}(n)), \mathbf{w}_Q >_{\mathbb{Q}} = \left[\eta \sum_{m=1}^{n} \left(\widehat{\Phi}(\mathbf{x}(m))e^*(m)\right)\right]^H \widehat{\mathbf{\Phi}}(\mathbf{x}(n))$$
$$= \eta \sum_{m=1}^{n} \left[e(m)\widehat{\mathbf{\Phi}}^H(\mathbf{x}(m))\widehat{\mathbf{\Phi}}(\mathbf{x}(n))\right]$$
$$= \eta \sum_{m=1}^{n} \left[e(m)\, \kappa_{\sigma,\mathbb{Q}^d}(\mathbf{x}(n),\ \mathbf{x}(m))\right] \qquad (10.57)$$

which can be seen to have the same form as Complex KLMS, aside from the quaternionic kernel used. We refer to this as the iQuaternion Kernel LMS (iQuat-KLMS). The next section focuses on simulations to verify the benefit of these approaches.

## 10.5 Simulation Results

We now evaluate the performance of the Quat-KLMS algorithm of Sect. 10.2 based on a simulated quaternion channel with nonlinearity. The channel structure follows [5], although extended to quaternions, and is based on the Weiner nonlinear model. It consisted of widely linear portion

$$z = g_1^* u(n) + g_2^* u^i(n) + g_3^* u^j(n) + g_4^* u^k(n)$$
$$+ h_1^* u(n-1) + h_2^* u^i(n-1) + h_3^* u^j(n-1) + h_4^* u^k(n-1) \qquad (10.58)$$

followed by the nonlinearity

$$y = z + az^2 + bz^3 \qquad (10.59)$$

where coefficients $g_1 \ldots g_4$, $h_1 \ldots h_4$, $a$, $b$ are quaternion-valued.

The coefficient values used were

$$g_1 = -0.45 - 0.4i + 0.3j + 0.15k, \quad h_1 = 0.15 + 0.175i - 0.025j + 0.1k$$
$$g_2 = 0.2 - 0.35i - 0.15j - 0.05k, \quad h_2 = -0.075 + 0.15i - 0.225j + 0.125k$$
$$g_3 = -0.05 - 0.1i - 0.4j + 0.2k, \quad h_3 = -0.05 + 0.025i + 0.075j - 0.05k$$
$$g_4 = -0.15 + 0.35i + 0.1j - 0.1k, \quad h_4 = 0.175 - 0.05i - 0.075j - 0.075k$$
$$a = -0.05 + 0.075i + 0.35j + 0.1k, \quad b = 0.03 - 0.025i - 0.25j + 0.05k$$
$$(10.60)$$

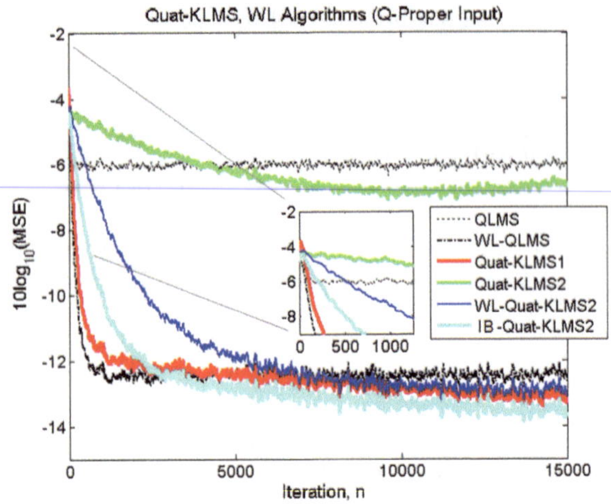


**Fig. 10.1** Quat-KLMS methods (In: $\mathbb{Q}$-proper, noise $= -17$ dBm)

For input sequence $u(n)$, the quaternion data used was either $\mathbb{Q}$-proper or $\mathbb{Q}$-improper. For $\mathbb{Q}$-proper, the data was

$$u(n) = 0.5u_a(n) + i0.5u_b(n) + j0.5u_c(n) + k0.5u_d(n) \tag{10.61}$$

and for $\mathbb{Q}$-improper:

$$u(n) = \sqrt{28/32}\,u_a(n) + i\sqrt{2/32}\,u_b(n) + j\sqrt{1/32}\,u_c(n) + k\sqrt{1/32}\,u_d(n) \tag{10.62}$$

where $u_a(n)$ to $u_d(n)$ are white Gaussian noise with unit variance. The channel output power was $-4.5$ dB with $\mathbb{Q}$-proper input, and $-2.8$ dB with $\mathbb{Q}$-improper input. The measurement noise was $\mathbb{Q}$-proper, and various noise levels were used.

This channel was tested with the following settings:

Figure 10.1: $\mathbb{Q}$-proper input, Meas. noise $= -17$ dBm.
Figure 10.2: $\mathbb{Q}$-improper input, Meas. noise $= -17$ dBm.
Figure 10.3: $\mathbb{Q}$-proper input, Meas. noise $= -40$ dBm.
Figure 10.4: $\mathbb{Q}$-improper input, Meas. noise $= -40$ dBm.
Figure 10.5: $\mathbb{Q}$-proper input, Meas. noise $= -12$ dBm.
Figure 10.6: $\mathbb{Q}$-improper input, Meas. noise $= -12$ dBm.

Figures 10.1 and 10.2 illustrate algorithm performance with both measurement noise and nonlinearity present. Figures 10.3 and 10.4 represent the use of a relatively noise-free channel. Finally, Figs. 10.5 and 10.6 show performance when the measurement noise is substantially larger compared to the nonlinearity.

With Figs. 10.1 and 10.2, the results are ensemble-averaged over 100 iterations. A 50-sample moving window average was also applied to the MSE before graphing, in order to provide more visible convergence curves. In the remaining figures (Figs. 10.3, 10.4, 10.5 and 10.6), ensemble-averaging over 25 iterations and

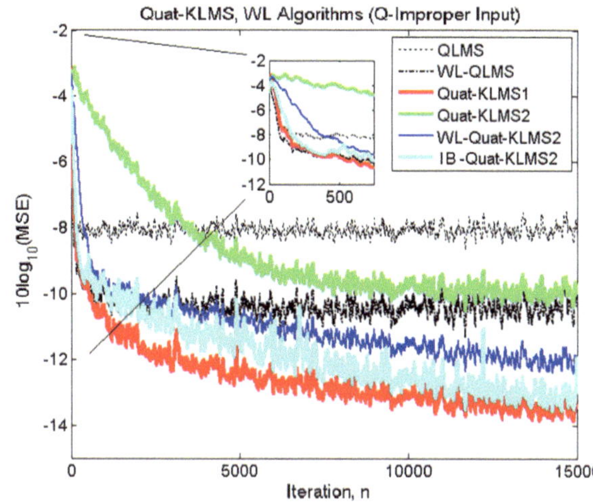


**Fig. 10.2** Quat-KLMS methods (In: $\mathbb{Q}$-improper, noise $= -17$ dBm)

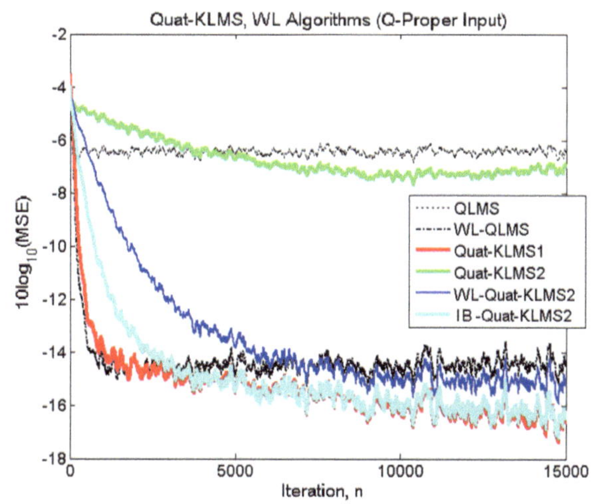


**Fig. 10.3** Quat-KLMS methods (In: $\mathbb{Q}$-proper, noise $= -40$ dBm)

100-sample moving window average were used. Figures 10.1 and 10.2 also contain an inserted graph to clarify the initial convergence behavior for each algorithm tested.

The algorithms and parameters used in Figs. 10.1, 10.2, 10.3, 10.4, 10.5 and 10.6 were

$$\text{QLMS: } \mu = 0.08$$

$$\text{WL-QLMS:} \mu = 0.02$$

$$\text{Quat-KLMS1: } \eta = 0.35, \ \sigma = \sqrt{1/8}$$

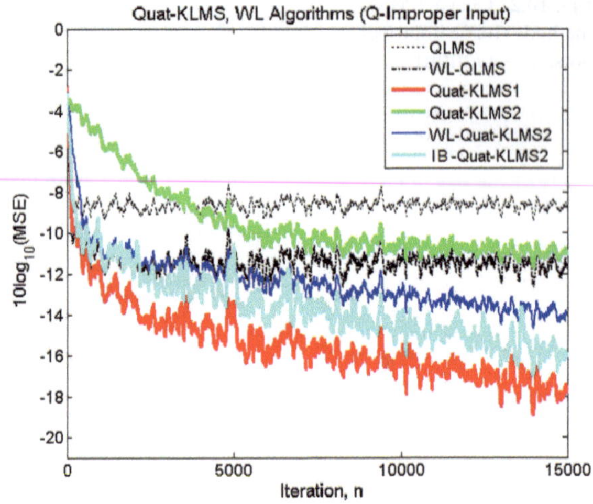


**Fig. 10.4** Quat-KLMS methods (In: $\mathbb{Q}$-improper, noise = −40 dBm)

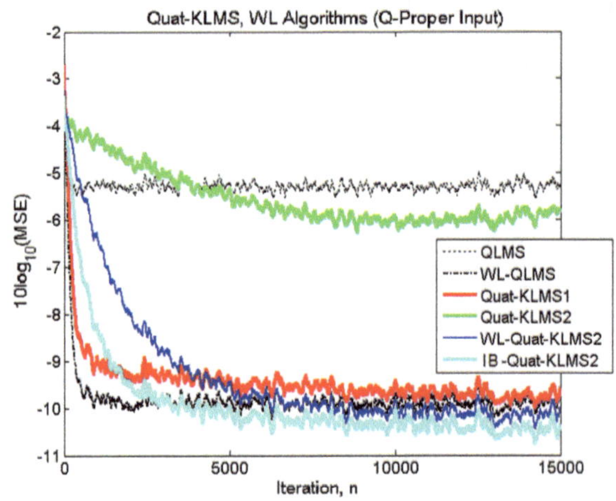


**Fig. 10.5** Quat-KLMS methods (In: $\mathbb{Q}$-proper, noise = −12 dBm)

$$\text{Quat-KLMS2: } \eta = (0.35/40), \;\; \sigma = \sqrt{1/16}$$

$$\text{WL-Quat-KLMS2: } \eta = (0.35/4), \;\; \sigma = \sqrt{1/16}$$

$$\text{IB-Quat-KLMS2: } \eta = (0.35/4), \;\; \sigma = \sqrt{1/16}$$

where KLMS1 denotes the use of the Quat-Ext Gaussian kernel, and Quat-KLMS2 the Quaternion Gaussian kernel. The parameters $\eta$, σ for Quat-KLMS were chosen based on exhaustive trial-and-error testing. For WL-Quat-KLMS, $\eta$ was chosen based on the use of WL estimation versus linear estimation, i.e., the effect on filter length.

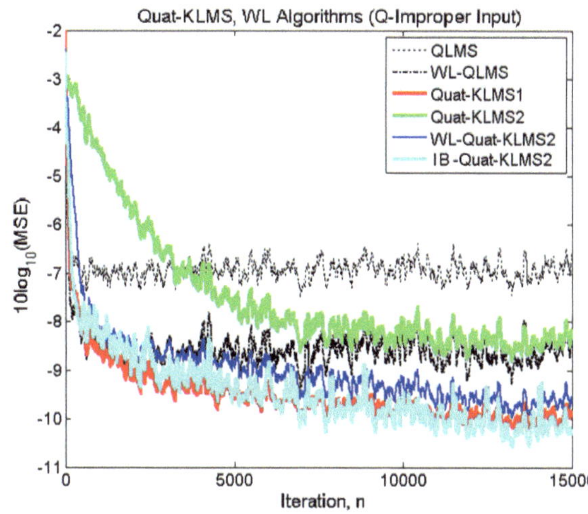


**Fig. 10.6** Quat-KLMS methods (In: $\mathbb{Q}$-improper, noise $= -12$ dBm)

The choice of $\eta$ for the IB-Quat-KLMS algorithm was similarly considered, while also considering how higher-order dimension terms diminish with $\sigma$ in Sect. 10.2.1. Methods for finding optimal values for $\sigma$ here are outside the scope of this work.

It should be noted that the step size $\eta$ for Quat-KLMS2 was particularly reduced based on observed stability issues. A bound for step-size $\eta$ was considered in Appendix F, however its evaluation is dependent on understanding the kernel transform space associated with each algorithm. This is needed in order to determine the eigenvalues of correlation matrix $\mathbf{R}$ for the transformed input. This was considered outside the scope here, and remains for future work.

Finally, sparsification was used in all tests performed for reduced complexity. Sparsification methods curb the linear growth of the kernel adaptive filter structure (see (10.46)–(10.48)) by choosing samples to discard, which should minimally affect performance. The results of all tests performed used the vector quantization (VQ) approach [13]. The sparsification parameters used and results obtained for MSE, network size after 15,000 iterations are shown in Table 10.1.

Note that the same sparsification parameter value was used for all algorithms, and only varied depending on whether the input was $\mathbb{Q}$-proper or $\mathbb{Q}$-improper. This

**Table 10.1** Sparsification parameter, network size after 15,000 iterations

| Test | Quantization size | Network size |
|---|---|---|
| Figure 1, All Algo. | 1.167 | 2088 |
| Figure 2, All Algo. | 0.475 | 2281 |
| Figure 3, All Algo. | 1.167 | 2088 |
| Figure 4, All Algo. | 0.475 | 2301 |
| Figure 5, All Algo. | 1.167 | 2103 |
| Figure 6, All Algo. | 0.475 | 2272 |

resulted in the same dictionary across algorithms, and was considered here to provide an unbiased algorithm comparison. Performance was seen to vary by less than 1 dB for all tests with sparsification.

From Figs. 10.1 and 10.2, we see Quat-KLMS1, WL-Quat-KLMS2 and IB-Quat-KLMS2 perform well for both $\mathbb{Q}$-proper, $\mathbb{Q}$-improper input, achieving $-12$ to $-13$ dBm with both distribution types. The algorithms provide an improvement on WL-QLMS of around 1.5 dB for $\mathbb{Q}$-proper input and 3 dB for $\mathbb{Q}$-improper.

On the other hand, Quat-KLMS2 has surprisingly worse performance (by 6 dB with $\mathbb{Q}$-proper, 3 dB with $\mathbb{Q}$-improper input). This should be due to the inability of the Quat-KLMS2 algorithm to perform well for channels which are a function of the perpendicular involutions of the input, where the use of QWL estimation may provide benefit. The associated RKHS space may also affect algorithm stability. Finally, in both cases, QLMS had a large MSE value (7 dB worse for $\mathbb{Q}$-proper input and 5 dB for $\mathbb{Q}$-improper input) compared to Quat-KLMS1.

Table 10.2 shows the steady-state MSE for each algorithm of Figs. 10.1 and 10.2. Both the average and standard deviation of the MSE are provided.

Figures 10.3 and 10.4 illustrate the performance for a relatively noise-free measurement channel. The performance of Quat-KLMS1 and the WL estimation based methods provides substantial improvement with both $\mathbb{Q}$-proper and $\mathbb{Q}$-improper input, yielding benefits of 3 dB and 5 dB, respectively. The results are otherwise similar to Figs. 10.1 and 10.2, as expected. Convergence is slow towards $-40$ dB, which may be attributed to the high-dimensional space used for learning, which affects both convergence rate and misadjustment.

Figures 10.5 and 10.6 show the MSE results with measurement noise at $-12$ dB. Here, the measurement noise is substantial compared to the performance degradation due to nonlinearity. This can be verified by comparing with the WL-QLMS of Figs. 10.3 and 10.4. We observe that the benefit of kernel-based methods here is not substantial. Specifically, the kernel methods provide less than 1 dB of benefit

**Table 10.2** MSE, standard dev. for various algorithms tested

| Algorithm | MSE $\pm$ standard deviation |
|---|---|
| QLMS | Figure 1: $0.24969 \pm 0.02749$ ($\mathbb{Q}$-proper)<br>Figure 2: $0.15679 \pm 0.03859$ ($\mathbb{Q}$-improper) |
| WL-QLMS | Figure 1: $0.05696 \pm 0.01051$ ($\mathbb{Q}$-proper)<br>Figure 2: $0.09121 \pm 0.03242$ ($\mathbb{Q}$-improper) |
| Quat-KLMS1 | Figure 1: $0.04888 \pm 0.00698$ ($\mathbb{Q}$-proper)<br>Figure 2: $0.04484 \pm 0.01033$ ($\mathbb{Q}$-improper) |
| Quat-KLMS2 | Figure 1: $0.21313 \pm 0.02226$ ($\mathbb{Q}$-proper)<br>Figure 2: $0.10138 \pm 0.02866$ ($\mathbb{Q}$-improper) |
| WL-Quat-KLMS2 | Figure 1: $0.05135 \pm 0.00910$ ($\mathbb{Q}$-proper)<br>Figure 2: $0.06304 \pm 0.01697$ ($\mathbb{Q}$-improper) |
| IB-Quat-KLMS2 | Figure 1: $0.04442 \pm 0.00745$ ($\mathbb{Q}$-proper)<br>Figure 2: $0.04968 \pm 0.02007$ ($\mathbb{Q}$-improper) |

with $\mathbb{Q}$-proper input, and 2 dB benefit with $\mathbb{Q}$-improper input. For the Quat-KLMS1, note that performance may benefit from a modified choice of kernel parameter $\sigma$. However, this was not done here in favor of using a uniform $\sigma$ for all tests. The results obtained indicate that the use of kernel methods should be considered only when the main source of channel distortion is due to nonlinearity.

A nonstationary channel test was also performed, which can be seen in Fig. 10.7. Specifically, a channel where an abrupt coefficient change occurs was tested. The test settings were the same used for Fig. 10.2 ($\mathbb{Q}$-improper data, $-17$ dBm noise). At iteration 7500, however, the coefficients were changed to

$$\begin{aligned} g_1 &= -0.45 - 0.4i + 0.3j + 0.15k, \quad h_1 = 0.15 + 0.175i - 0.025j + 0.1k \\ g_2 &= 0.2 - 0.35i - 0.15j - 0.05k, \quad h_2 = -0.075 + 0.15i - 0.225j + 0.125k \\ g_3 &= -0.05 - 0.1i - 0.4j + 0.2k, \quad h_3 = -0.05 + 0.025i + 0.075j - 0.05k \\ g_4 &= -0.15 + 0.35i + 0.1j - 0.1k, \quad h_4 = 0.175 - 0.05i - 0.075j - 0.075k \\ a &= -0.05 + 0.075i + 0.35j + 0.1k, \quad b = 0.03 - 0.025i - 0.25j + 0.05k \end{aligned} \tag{10.63}$$

In addition, only 25 iterations were used for the ensemble averaging, and a 100-sample moving window average was applied to the MSE with Fig. 10.7.

From Fig. 10.7, we see that the convergence is slower for the kernel-based methods, which is expected. However, the convergence behavior is reasonably quick, as well as stable. This should be partially attributed to the vector quantization sparsification method, though the results also indicate the robustness of the quaternion kernel methods described. The MSE performance of all methods is essentially as expected, with Quat-KLMS1 and WL methods yielding benefit.

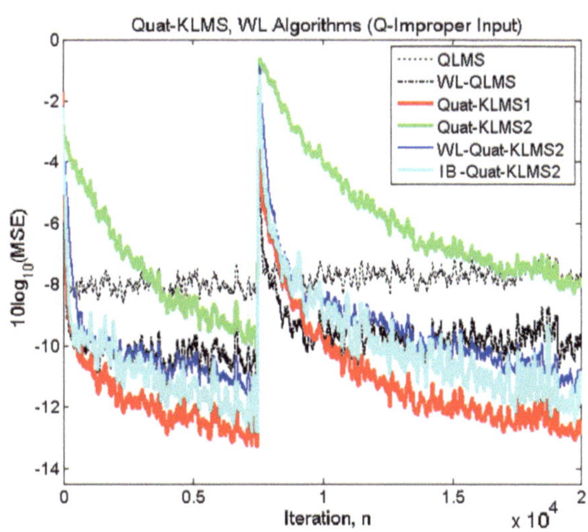


**Fig. 10.7** Quat-KLMS, WL methods with nonstationary channel

Finally, Fig. 10.8 shows the use of these quaternion-based algorithms in an application using electroencephalogram (EEG) data. The test data used is summarized here, and more details can be found from reference [14].

The EEG data was collected from electrodes placed at various locations on the head, based on the 10–20 system. For each case, eight collected signals were used:

FP1, FP2—frontal electrodes

C3, C4—central electrodes

O1, O2—occipital electrodes

vEOG, hEOG—EOG measurements

The data used was data collected for the scenario where an eyebrow was raised.

For forming quaternion input, the signals were grouped based on pattern similarity (i.e. FP1 and FP2 are symmetrically located, signals C3 and C4, etc.), and the eight signals were considered a tuple of quaternion inputs.

Specifically, the 8 input signals are divided into two consecutive quaternion inputs: $(\mathrm{FP1} + i\mathrm{C3} + j\mathrm{O1} + k\mathrm{vEOG})$ and $(\mathrm{FP2} + i\mathrm{C4} + j\mathrm{O2} + k\mathrm{hEOG})$. This formed the input of a one-step predictor of order $L = 10$, based on the various algorithms. For the tests here, the parameters used were

$$\text{QLMS: } \mu = 0.08$$

$$\text{WL-QLMS: } \mu = 0.02$$

$$\text{Quat-KLMS1: } \eta = 0.6,\ \sigma = \sqrt{1/8}$$

$$\text{Quat-KLMS2: } \eta = (0.6/40), \sigma = \sqrt{1/16}$$

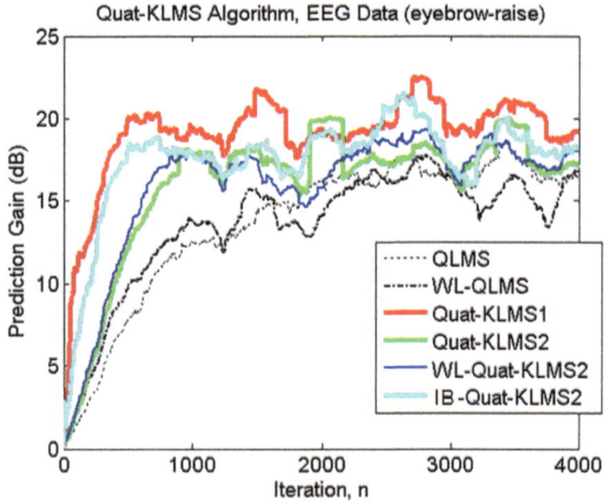


**Fig. 10.8** Quat-KLMS, WL methods with EEG data (prediction task)

$$\text{WL-Quat-KLMS2: } \eta = (0.6/4), \ \sigma = \sqrt{1/16}$$

$$\text{IB-Quat-KLMS2: } \eta = (0.6/4), \ \sigma = \sqrt{1/16}$$

where $\sigma_{in}^2$ is average input power. The prediction gain performance is graphed for each algorithm.

From Fig. 10.8, we see that the Quat-KLMS algorithms (particularly Quat-KLMS1 and IB-Quat-KLMS2) provide improved steady-state prediction gain, as expected. QLMS shows worse prediction gain than WL-QLMS (around 2 dB worse). The convergence rate of the kernel-based algorithms is also improved (reaches steady-state in less than 500 iterations). The benefit of Quat-KLMS, WL versions, may be attributed to any nonlinearity in the EEG data, in how the current sample is related to past samples.

The simulations of Figs. 10.1, 10.2, 10.3, 10.4, 10.5, 10.6, 10.7 and 10.8 show the ability of the quaternion kernel-based algorithms in learning nonlinear channels, as well as the benefit of the Quat-Ext Gaussian kernel and QWL estimation in improving performance with quaternion data. Note that Quat-KLMS1 performs equally well with the WL-based algorithms, which can be understood based on the Sect. 10.2 description.

For illustrating iQuat-KLMS performance, graphs of both iQuat-KLMS and Quat-KLMS for the synthetic channel described previously (with noise at $-17$ dBm) are shown in Figs. 10.9 and 10.10. For both, the parameters used were $\eta = 0.35$ and $\sigma = \sqrt{1/8}$, which were found to yield the best performance.

For both iQuat-KLMS and Quat-KLMS, the MSE reached $-14$ dBm for both $\mathbb{Q}$-proper and $\mathbb{Q}$-improper data. However, an improvement in convergence can be seen based on the use of iQuat-KLMS versus Quat-KLMS (converges 500–1000 iterations sooner), while the steady-state MSE remains the same. This is consistent with the

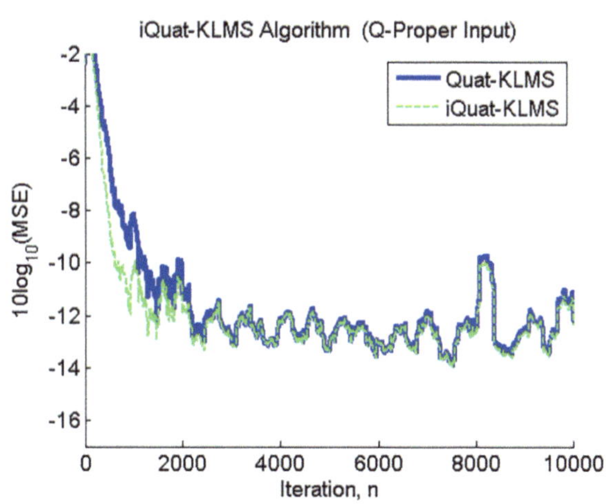


**Fig. 10.9** iQuat-KLMS with Q-proper data

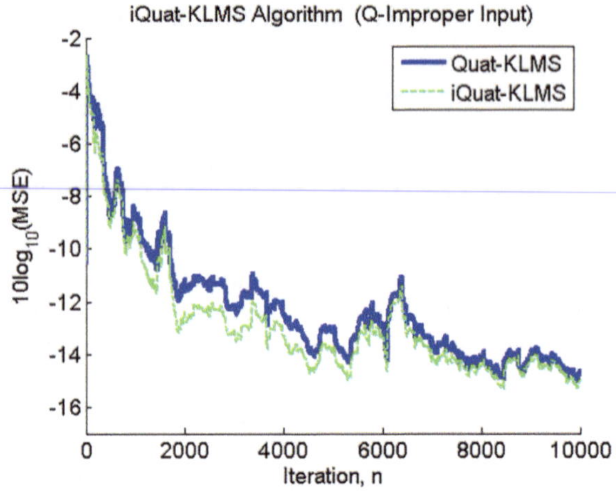


**Fig. 10.10** iQuat-KLMS with Q-improper data

discussion from Chap. 5 on the benefit of i-gradient for improving the convergence steepness with real cost functions.

## 10.6 Conclusion

Here we described the derivation of a kernel adaptive algorithm with quaternion data, which we refer to as the Quaternion KLMS (Quat-KLMS) and iQuaternion KLMS (iQuat-KLMS). The approach follows that of Bouboulis and Theodoridis for Complex KLMS [15], and involves describing suitable kernel functions with quaternion data and the associated RKHS spaces. Also, a description of a modified $\mathbb{HR}$ calculus to perform gradient operations on a quaternion RKHS was provided. Simulations of a nonlinear quaternion channel under various SNRs, as well as a nonstationary channel test, were used to verify the potential benefits of Quat-KLMS to robustly handle channel nonlinearities. The convergence benefit of iQuat-KLMS, based on i-gradient, was also observed.

## References

1. E.J.B. Corrochano, N.A. Daniel, Clifford support vector machines for classification, regression, and recurrence. IEEE Trans. Neural Netw. **21**(11), 1731–1746 (2010)
2. A. Shilton, Mercer's Theorem for Quaternionic Kernels, Univ. Melbourne, Dept. Elect. Electron. Eng., Melbourne, Australia, Tech. Rep. TR07-009. http://www.ee.unimelb.edu.au/staff/apsh/publications/
3. A. Shilton, D.T.H. Lai, Quaternionic and complex-valued support vector regression for equalization and function approximation, in *Proc. of Inter. Joint Conf. on Neural Networks* (2007)

4. D.P. Mandic, C. Jahanchahi, C. Cheong Took, A quaternion gradient operator and its applications. IEEE Signal Process. Lett. **18**(1), 47–50 (2011)
5. P. Bouboulis, S. Theodoridis, Extension of Wirtinger's calculus to reproducing Kernel Hilbert spaces and the complex Kernel LMS. IEEE Trans. Signal Process. **59**(3) (2011)
6. W. Liu, J. Principe, S. Haykin, *Kernel Adaptive Filtering: A Comprehensive Introduction* (Wiley Publishers, 2010)
7. A.W. Knapp, *Basic Algebra* (Springer Publishers, 2006)
8. I. Steinwart, D. Hush, C. Scovel, An explicit description of the reproducing Kernel Hilbert spaces of Gaussian RBF kernels. IEEE Trans. Inf. Theory **52**(10) (2006)
9. S. Sarkka, Notes on Quaternions, https://www.scribd.com/document/357419773/Notes-on-Quaternions-Simo-Sarkka
10. J. Mathews, G. Sicuranza, *Polynomial Signal Processing* (Wiley Publishers, 2000)
11. B. Picinbono, Bernard, Chevalier, Pascal, Widely linear estimation with complex data. IEEE Trans. Signal Process. **43**(8), 2030–2033 (1995)
12. P. Bouboulis, S. Theodoridis, M. Mavroforakis, The augmented complex Kernel LMS. IEEE Trans. Signal Process. **60**(9), 4962–4967 (2012)
13. B. Chen, S. Zhao, P. Zhu, J.C. Principe, Quantized kernel least mean square algorithm. IEEE Trans. Neural Netw. Learn. Syst. **23**(1), 22–32 (2012)
14. S. Javidi, C. Cheong Took, C. Jahanchahi, N. Le Bihan, and D. P. Mandic, "Blind Extraction of Improper Quaternion Sources", IEEE Proceedings of Inter. Conf. on Acoustics, Speech and Signal Processing (ICASSP), May 2011, pp. 3078–3711. http://www.commsp.ee.ic.ac.uk/~mandic/research
15. S.L. Gay, J. Benesty, *Acoustic Signal Processing for Telecommunication* (Kluwer, Boston, MA, 2000)

4. D.P. Mandic, C. Jahanchahi, C. Cheong Took, A quaternion gradient operator and its applications. IEEE Signal Process. Lett. 18, 47–50 (2011)
5. P. Bouboulis, S. Theodoridis, Extension of Wirtinger's calculus to reproducing Kernel Hilbert spaces and the complex kernel LMS. IEEE Trans. Signal Process. 59 (2011)
6. W. Liu, J. Príncipe, S. Haykin, *Kernel Adaptive Filtering: A Comprehensive Introduction* (Wiley Publishing, 2010)

[illegible]

# Chapter 11
# The Quaternion Kernel Maximum Correntropy Algorithm

This chapter describes a kernel adaptive filter for quaternion data based on maximizing correntropy. The Kernel Maximum Correntropy method uses correntropy to measure similarity between the filter output and desired response (see Chap. 6). The approach improves performance for biased or non-Gaussian signals when compared to the minimum MSE criterion of Kernel LMS.

Here, we apply the modified $\mathbb{HR}$ calculus for Hilbert spaces from Chap. 10 for evaluating the cost function gradient. The result is referred to as the Quaternion Kernel Maximum Correntropy (Quat-KMC) algorithm, and is shown using simulations to improve performance with non-Gaussian inputs.

## 11.1 Goal of the Algorithm

For understanding the potential benefit of maximizing correntropy, the system goal considering quaternion data is outlined here. The system goal is to form the function $f$, based on sequence $(\mathbf{u}_1, d_1)$, $(\mathbf{u}_2, d_2)$, ..., $(\mathbf{u}_N, d_N)$, where $\mathbf{u}_i$ is the system input, and $d_i$ is the desired response, both quaternion-valued, for sample time $i$. The desired response can be expressed: $d_i = f_{\text{opt}}(\mathbf{u}_i) + \xi_i$, where $f_{\text{opt}}$ is some unknown nonlinear quaternion function to be estimated, and $\xi_i$ is noise with an arbitrary pdf (not necessarily Gaussian).

When finding $f$, we ideally would like to solve the empirical risk minimization (ERM) problem, i.e. minimize

$$\text{R}_{\text{emp}}[f \in Q,\ \mathbf{Z}_N] = \sum_{i=1}^{N} \left|f_{\text{opt}}(\mathbf{u}_i) - f(\mathbf{u}_i)\right|^2 \tag{11.1}$$

where $Q$ is a quaternion vector space (becomes a quaternion RKHS when using a quaternionic kernel). The use of MSE can result in large variations for the



T. K. Paul and T. Ogunfunmi, *Adaptive Filter Algorithms For Complex and Quaternion Data*, https://doi.org/10.1007/978-3-032-23798-9_11

weights, or shift in output, when the noise distribution contains outlying points, is non-symmetric, or has a non-zero mean.

The definition of correntropy is expressed (see Chap. 4):

$$V_\sigma(X, Y) = E[\kappa_\sigma(X - Y)] \tag{11.2}$$

where $X$ and $Y$ are quaternion random variables, and $\kappa_\sigma(.)$ is a symmetric positive definite quaternionic kernel, with kernel size $\sigma$. The parameter $\sigma$ adjusts the mapping where similarity is measured. Note that the kernel function used is required to be symmetric.

Based on real inputs, the Gaussian kernel can be expressed

$$\kappa_{\text{real},\sigma}(X - Y) = \frac{1}{\sqrt{2\pi}\,\sigma} \exp\left[\frac{-1}{2\sigma^2}(X - Y)^2\right]$$

where $X$, $Y$ are real-valued. A similar Gaussian-based kernel for quaternion data may be expressed as

$$\begin{aligned}\kappa_\sigma(X - Y) &= \frac{4}{\sqrt{2\pi\sigma}} \exp\left[\frac{-1}{2\sigma^2}((X_a - Y_a)^2 + (X_b - Y_b)^2\right.\\ &\left.\quad +(X_c - Y_c)^2 + (X_d - Y_d)^2)\right]\\ &= \frac{4}{\sqrt{2\pi\sigma}} \exp\left[\frac{-1}{2\sigma^2}|X - Y|^2\right]\end{aligned} \tag{11.3}$$

where $X$ and $Y$ are now quaternion-valued, of the form: $X = X_a + iX_b + jX_c + kX_d$ and $Y = Y_a + iY_b + jY_c + kY_d$. A description of this kernel can be found in Sect. 10.2.

For estimating correntropy (11.2), the expectation may be approximated based on a finite set of input data using

$$\hat{V}_{N,\sigma}(X, Y) = \frac{1}{N}\sum_{i=1}^{N} \kappa_\sigma(X - Y)$$

The correntropy properties using real data should similarly apply to quaternion data. In particular, the correntropy should contain all even moments of the difference between $X$ and $Y$ (specifically the magnitude $|X - Y|$ where $X$ and $Y$ are quaternion-valued). Also, the higher-order terms should decay with increasing $\sigma$, such that the second-order term dominates with large $\sigma$. This should allow a gradient-based approach to be valid for minimizing correntropy (properties 2, 3 of Sect. 6.3.1).

For maximizing correntropy, we can form a cost function

$$J_n = \frac{1}{N} \sum_{i=n-N+1}^{n} \kappa_\sigma(d_i - y_i) \tag{11.4}$$

where $d_i$ is the desired response, $y_i$ is the adaptive filter output and $N$ is the number of samples for the correntropy estimate. The objective is maximizing the correntropy (i.e. similarity measure) between $d_i$ and $y_i$.

## 11.2 Derivation of the Algorithm

For the kernel maximum correntropy approach with quaternion data, we maximize (11.4) for a kernel filter. The filter can be expressed $y_n =< \Phi(\mathbf{u}_n), \mathbf{w}_n >$ which may also be written

$$y_n = \mathbf{w}_n^H \varphi_n \tag{11.5}$$

where $\phi_n = \Phi(\mathbf{u}_n)$ and $\Phi(.)$ is the kernel map to a quaternion RKHS (see Appendix E). Note the kernel size used for $\Phi(\mathbf{u}_n)$ differs from $\sigma$ of the correntropy (11.4), in general. The kernel size for $\Phi(\mathbf{u}_n)$ will be denoted $\sigma_\mathbf{u}$ when applicable.

Maximizing (11.4) is performed using gradient ascent, since the goal is maximizing versus minimizing cost. Using $N = 1$ in (11.4), i.e. instantaneous correntropy estimate:

$$\begin{aligned}
\mathbf{w}_{n+1} &= \mathbf{w}_n + \eta \frac{\partial}{\partial \mathbf{w}_n^*}[J_n] \\
&= \mathbf{w}_n + \eta \frac{\partial}{\partial \mathbf{w}_n^*}[\kappa_\sigma(d_n, y_n)] \\
&= \mathbf{w}_n + \eta \frac{\partial}{\partial \mathbf{w}_n^*}\left[\frac{4}{\sqrt{2\pi}\,\sigma}\exp\left[\frac{-|d_n - y_n|^2}{2\sigma^2}\right]\right] \\
&= \mathbf{w}_n + \eta \frac{\partial}{\partial \mathbf{w}_n^*}\left[\exp\left[\frac{-\left|d_n - \mathbf{w}_n^H \varphi_n\right|^2}{2\sigma^2}\right]\right]
\end{aligned} \tag{11.6}$$

The scaling factor $4/(\sqrt{2\pi}\,\sigma)$ is included in $\eta$ in the final step.

For the gradient, the function to be differentiated is defined on a quaternion RKHS, and has a real-valued output. For this case, the modified $\mathbb{HR}$ calculus applies.

The result can be seen to be

$$\begin{aligned}
\frac{\partial}{\partial \mathbf{w}_n^*}\left[\exp\left[\frac{-\left|d_n - \mathbf{w}_n^H \varphi_n\right|^2}{2\sigma^2}\right]\right] &= \left[\varphi_n \frac{\left(d_n - \mathbf{w}_n^H \varphi_n\right)^*}{2\sigma^2} - \frac{\left(d_n - \mathbf{w}_n^H \varphi_n\right)}{4\sigma^2}\varphi_n^*\right] \\
&\quad \exp\left[\frac{-\left|d_n - \mathbf{w}_n^H \varphi_n\right|^2}{2\sigma^2}\right] \\
&= \frac{1}{2\sigma^2}\left[\varphi_n e_n^* - \frac{1}{2} e_n \varphi_n^*\right]\exp\left[\frac{-|e_n|^2}{2\sigma^2}\right]
\end{aligned} \tag{11.7}$$

where $e_n = \left(d_n - \mathbf{w}_n^H \phi_n\right)$ is the estimation error. Note that multiplication between two quaternion values is non-commutative, while multiplication between a quaternion value and a real scalar is commutative.

The weight update may be expressed as

$$\mathbf{w}_{n+1} = \mathbf{w}_n + \eta \exp\left[\frac{-|e_n|^2}{2\sigma^2}\right]\left[\varphi_n e_n^* - \tfrac{1}{2} e_n \varphi_n^*\right] \tag{11.8}$$

with scaling $1/2\sigma^2$ again in $\eta$. Using (11.8) iteratively with $\mathbf{w}_0 = 0$

$$\mathbf{w}_1 = \eta \exp\left[\frac{-|e_0|^2}{2\sigma^2}\right]\left[\varphi_0 e_0^* - \tfrac{1}{2} e_0 \varphi_0^*\right]$$

$$\mathbf{w}_2 = \eta \exp\left[\frac{-|e_0|^2}{2\sigma^2}\right]\left[\varphi_0 e_0^* - \tfrac{1}{2} e_0 \varphi_0^*\right] + \eta \exp\left[\frac{-|e_1|^2}{2\sigma^2}\right]\left[\varphi_1 e_1^* - \tfrac{1}{2} e_1 \varphi_1^*\right]$$

$$\vdots$$

$$\mathbf{w}_{n+1} = \eta \sum_{i=0}^{n}\left[\exp\left[\frac{-|e_i|^2}{2\sigma^2}\right]\left[\varphi_i e_i^* - \tfrac{1}{2} e_i \varphi_i^*\right]\right] \tag{11.9}$$

Equation (11.9) shows the filter weights for the Quaternion KMC algorithm (referred to here as Quat-KMC for short).

For the Quat-KMC filter output (i.e. estimates), we have

$$\begin{aligned} y_n =< \Phi(\mathbf{u}_n), \mathbf{w}_n > &= \eta \sum_{i=0}^{n-1}\left[\exp\left[\frac{-|e_i|^2}{2\sigma^2}\right]\left[\varphi_i e_i^* - \tfrac{1}{2} e_i \varphi_i^*\right]^H \varphi_n\right] \\ &= \eta \sum_{i=0}^{n-1}\left[\exp\left[\frac{-|e_i|^2}{2\sigma^2}\right]\left[e_i \varphi_i^H - \tfrac{1}{2} \varphi_i^T e_i^*\right]\varphi_n\right] \end{aligned} \tag{11.10}$$

For use of the 'kernel trick', we follow the approach of Chap. 10 used for deriving Quat-KLMS, i.e. expressing the RKHS in terms of a real RKHS map and quaternion scalar $s_\Phi$.

We can then write

$$\begin{aligned} y_n &=< \Phi(\mathbf{u}_n), \mathbf{w}_{n-1} > \\ &= \eta \sum_{i=0}^{n-1}\left[\exp\left[\frac{-|e_i|^2}{2\sigma^2}\right]\left[e_i \Phi_i^T s_\Phi^* - \tfrac{1}{2} \Phi_i^T s_\Phi e_i^*\right] s_\Phi \Phi_n\right] \\ &= \eta \sum_{i=0}^{n-1}\left[\exp\left[\frac{-|e_i|^2}{2\sigma^2}\right]\left[e_i |s_\Phi|^2 - \tfrac{1}{2} s_\Phi e_i^* s_\Phi\right]\Phi_i^T \Phi_n\right] \end{aligned}$$

$$= \eta \sum_{i=0}^{n-1} \left[ \exp\left[ \frac{-|e_i|^2}{2\sigma^2} \right] \left[ e_i |s_\Phi|^2 - \tfrac{1}{2} s_\Phi e_i^* s_\Phi \right] \kappa_{\mathbb{R},\sigma_{\mathbf{u}}}(\mathbf{u}_{i,\,comp}, \mathbf{u}_{n,\,comp}) \right] \quad (11.11)$$

where $\Phi_n$ is the mapping $\Phi(\mathbf{u}_{n,\,comp})$ based on Sect. 10.2.2. This is the form of Quat-KMC implemented. A brief convergence analysis is shown in Appendix F.

## 11.3 Simulation Results

Here we evaluate Quat-KMC using simulations. The performance versus Quat-KLMS is shown using a nonlinear channel with non-Gaussian noise.

The channel consisted of the filter

$$\begin{aligned} z(n) &= g_1^* u(n) + g_2^* u^i(n) + g_3^* u^j(n) + g_4^* u^k(n) \\ &\quad + h_1^* u(n-1) + h_2^* u^i(n-1) + h_3^* u^j(n-1) + h_4^* u^k(n-1) \end{aligned}$$

and added non-Gaussian noise

$$y(n) = z(n) + az^2(n) + bz^3(n) + v(n)$$

where coefficients $g_1 \ldots g_4$, $h_1 \ldots h_4$, $a$, $b$ and added noise $v(n)$ are all quaternion-valued.

For the tests, both input $u(n)$, noise $v(n)$, were formed using impulsive Gaussian mixture models to form non-Gaussian signals. Specifically, the probability distributions used were

$$\begin{aligned} p_u(i) = &\,(0.85N(1.0, 0.01) + 0.15N(3.0, 0.01)) + i\,(0.40N(0.5, 0.01) \\ &+ 0.60N(2.5, 0.01)) + j(0.65N(3.5, 0.01) + 0.35N(1.5, 0.01)) \\ &+ k(0.25N(2.0, 0.01) + 0.75N(5.5, 0.01)) \end{aligned}$$

$$\begin{aligned} p_v(i) = &\,(0.90N(0.0, 0.01) + 0.10N(1.0, 0.01)) + i\,(0.70N(3.0, 0.01) \\ &+ 0.30N(0.5, 0.01)) + j(0.45N(1.0, 0.01) + 0.55N(4.5, 0.01)) \\ &+ k(0.80N(0.5, 0.01) + 0.20N(1.5, 0.01)) \end{aligned}$$

where $N(m_N, \sigma_N)$ denotes the normal (Gaussian) pdf with mean $m_N$, variance $\sigma_N$.

A graph of Quat-KMC and Quat-KLMS for the simulated channel described is shown in Fig. 11.1. For both the step-size, kernel sizes were chosen to minimize error. For Quat-KMC, parameters were $\eta = 0.35$, $\sigma_{\mathrm{u}} = 2.24$, $\sigma = 0.134$, while for Quat-KLMS $\eta = 0.35$ and $\sigma_{\mathrm{u}} = 2.24$ were used. The results show improvement of Quat-KMC modeling channel output when the input, noise are non-Gaussian compared to Quat-KLMS.

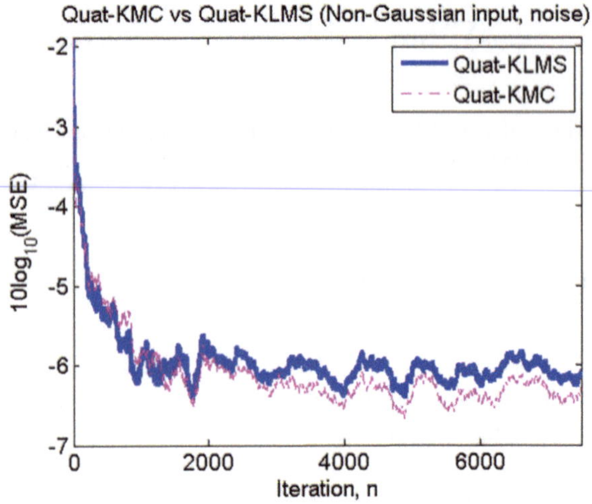


**Fig. 11.1** Quat-KMC, QuatKLMS for non-Gaussian signal

Note that Quat-KLMS still performs well (Quat-KMS benefit around 0.5 dB). This may be due to less deviation from Gaussianity of the quaternion mixture model form, or a benefit of the quaternion-based learning structure for modeling signals here.

## 11.4 Conclusion

This chapter showed the derivation of a kernel adaptive algorithm for quaternions based on maximizing the correntropy described in Chap. 6. The resulting algorithm is the Quaternion Kernel Maximum Correntropy (Quat-KMC) algorithm. Use of the modified $\mathbb{HR}$ calculus for Hilbert spaces from Chap. 10 allowed the gradient operation on a quaternion RKHS to be found easily. Simulation results showed the benefit of Quat-KMC for handling non-Gaussian signals versus Quat-KLMS.

# Appendix A
# Associated RKHS of the Real Gaussian Kernel

The following describes the associated reproducing kernel Hilbert space (RKHS) of the real Gaussian kernel. This RKHS is described in detail, particularly its properties, in Ref. [1].

The real Gaussian kernel function may be expressed as

$$\kappa(\mathbf{a}, \mathbf{b}) = \exp\left(-\sigma \sum_{i=1}^{d} (a_i - b_i)^2\right) \tag{A.1}$$

Initially, we consider the scalar input case. We can write

$$\begin{aligned}\kappa(a, b) &= \exp\left(-\sigma (a-b)^2\right)\\ &= \exp\left(-\sigma a^2 + \sigma 2ab - \sigma b^2\right)\\ &= e^{-\sigma a^2} e^{\sigma 2ab} e^{-\sigma b^2}\end{aligned} \tag{A.2}$$

where either $\exp(x)$ or $e^x$ is used to denote the exponential function.

Using the Taylor series expansion of $e^x$, i.e.

$$e^x = \sum_{n=0}^{\infty} \frac{x^n}{n!}$$

we obtain

$$\begin{aligned}\kappa(a, b) &= e^{-\sigma a^2} \left(\sum_{n=0}^{\infty} \frac{(2\sigma ab)^n}{n!}\right) e^{-\sigma b^2}\\ &= \sum_{n=0}^{\infty} \left(\sqrt{\frac{(2\sigma)^n}{n!}} a^n e^{-\sigma a^2} \cdot \sqrt{\frac{(2\sigma)^n}{n!}} b^n e^{-\sigma b^2}\right)\end{aligned} \tag{A.3}$$

which may be written in the form:



T. K. Paul and T. Ogunfunmi, *Adaptive Filter Algorithms For Complex and Quaternion Data*, https://doi.org/10.1007/978-3-032-23798-9

$$\kappa(a, b) = \varphi^T(b)\varphi(a) \tag{A.4}$$

where

$$\varphi(x) = e^{-\sigma x^2}\left[1 \quad \sqrt{2\sigma}\, x \quad \sqrt{\frac{(2\sigma)^2}{2!}}\, x^2 \quad \sqrt{\frac{(2\sigma)^3}{3!}}\, x^3 \quad \cdots\right]$$

Thus, kernel function (A.2) can be seen to be the associated inner product for the space generated with the transform map $\varphi(x)$. The representation in (A.4) is known as the Mercer condition, and is satisfied by any valid kernel function [2, 3]. It can be seen that the space generated by $\varphi(x)$ is infinite-dimensional, and may be used for learning polynomials of any order, where $\sigma$ controls the emphasis of the higher-order terms.

With vector inputs for the kernel function (A.1), the RHKS can be described as follows. Writing the dimensions of the scalar case as

$$e_n(x) = \sqrt{\frac{(2\sigma)^n}{n!}}\, x^n e^{-\sigma x^2} \tag{A.5}$$

for the vector case, an orthonormal basis for the RKHS was shown to be [1]

$$(e_{n_1} \otimes \cdots \otimes e_{n_d})_{n_1,\ldots,n_d \geq 0} \tag{A.6}$$

where the tensor product function, $f \otimes g := X \times X \to \mathbb{K}$, of functions $f, g := X \to \mathbb{K}$, is defined as

$$f \otimes g(x, x') := f(x)g(x'), \quad x, x' \in X \tag{A.7}$$

and the d-fold tensor product is defined similarly, i.e.:

$$f_1 \otimes f_2 \otimes \cdots f_d(x_1, \ldots, x_d) := f_1(x_1) f_2(x_2) \cdots f_d(x_d), \quad x_1, \ldots, x_d \in X \tag{A.8}$$

Using this basis, the inner product is

$$\begin{aligned}
\kappa(\mathbf{a}, \mathbf{b}) &= \sum_{n_1,\ldots,n_d=0}^{\infty} \left[e_{n_1} \otimes e_{n_2} \cdots \otimes e_{n_d}(\mathbf{a})\right]\left[e_{n_1} \otimes e_{n_2} \cdots \otimes e_{n_d}(\mathbf{b})\right] \\
&= \sum_{n_1,\ldots,n_d=0}^{\infty} \prod_{j=1}^{d} \frac{(2\sigma^2)^{n_j}}{n_j!} (a_j b_j')^{n_j} e^{-\sigma^2 a_j^2 - \sigma^2 b_j^2} \\
&= \prod_{j=1}^{d} \sum_{n_1,\ldots,n_d=0}^{\infty} \frac{(2\sigma^2)^{n_j}}{n_j!} (a_j b_j')^{n_j} e^{-\sigma^2 a_j^2 - \sigma^2 b_j^2}
\end{aligned}$$

$$= \prod_{j=1}^{d} e^{2\sigma^2 a_j b_j'} e^{-\sigma^2 a_j^2 - \sigma^2 b_j^2}$$

$$= e^{-\sigma^2 \sum_{j=0}^{d} (a_j - b_j)^2} \tag{A.9}$$

The complex Gaussian kernel has the same basis (A.6), although inputs **a**, **b** can be complex, and a complex inner product definition is used for (A.9) (see [1]).

# Appendix B
# Computation, Performance Tradeoff of CKLMS Versus CLMS

The following compares the computational cost of the Complex Kernel LMS (CKLMS) algorithm with the Complex LMS (CLMS), as well as MSE performance. The goal is to provide insight into the trade off between computation and performance for the algorithms.

For the Complex LMS, the following computations occur each iteration:

$$\begin{array}{ll} \hat{d}(n) = \mathbf{w}(n)^H \mathbf{x}(n) & (M \text{ mult.},\ \ M - 1 \text{ adds}) \\ \mathbf{w}(n+1) = \mathbf{w}(n) + \mu e^*(n)\mathbf{x}(n) & (M + 1 \text{ mult.},\ \ M + 1 \text{ adds}) \end{array} \tag{B.1}$$

The Complex LMS requires $2M + 1$ complex multiplications and $2M$ complex additions, where $M$ is the length of the filter, $\mathbf{w}(n)$. The choice of $M$ reflects the expected length of the channel, in samples.

For Complex Kernel LMS (CKLMS), each computation iteration is

$$\begin{aligned} \hat{d}(n) &= < \boldsymbol{\Phi}(\mathbf{z}(n)), \mathbf{w}(n-1) >_S \\ &= \mu \sum_{k=1}^{n-1} e(k) < \boldsymbol{\Phi}(\mathbf{z}(n)), \boldsymbol{\Phi}(\mathbf{z}(k)) >_S \end{aligned} \tag{B.2}$$

The CKLMS requires $n$' kernel evaluations, $n' + 1$ complex multiplications and $n'$ complex additions (where $n' = (n-1)$, and $n$ is the iteration number). The complexity of each kernel evaluation depends on the chosen kernel.

For example, with the polynomial kernel:

$$< \boldsymbol{\Phi}(\mathbf{z}(n)), \boldsymbol{\Phi}(\mathbf{z}(k)) >_S = (1 + \mathbf{z}(k)^H \mathbf{z}(n))^q$$

each kernel evaluation takes $M$ complex multiplications, $M$ complex additions and 1 exponent operation. Thus, the total computation for CKLMS in this case is: $(M + 1)n' + 1$ complex multiplications, $(M + 1)n'$ complex additions and $n'$ exponent operations.



T. K. Paul and T. Ogunfunmi, *Adaptive Filter Algorithms For Complex and Quaternion Data*, https://doi.org/10.1007/978-3-032-23798-9

For the complex Gaussian kernel

$$< \boldsymbol{\Phi}(\mathbf{z}(n)), \boldsymbol{\Phi}(\mathbf{z}(k)) >_S = \exp\left(-\sigma \sum_{i=1}^{M} (\mathbf{z}_i(n) - \mathbf{z}_i^*(k))^2\right)$$

each kernel evaluation takes $M + 1$ complex multiplications, $2M - 1$ complex additions and 1 exponential function evaluation. Thus, the CKLMS would take: $(M + 2)n' + 1$ complex multiplications, $2Mn'$ complex additions and $n'$ exponential function evaluations.

Summarizing the computational complexity of each (per iteration):

| | |
|---|---|
| Complex LMS: | 2M + 1 mult's, 2M adds |
| Complex KLMS (Poly.): | (M + 1)n′ + 1 mult's, (M + 1)n′ adds, n′ exponent eval's |
| Complex KLMS (Gauss.): | (M + 2)n′ + 1 mult's, 2Mn′ adds, n′ exp(.) eval's |

Thus, there is a substantial increase in computation with CKLMS, which is mostly based on the number of iterations, $n$. However, sparsification may be used, which reduces complexity (i.e. the range of index $k$ for (B.2) changes to $k = 1 \ldots L$, where only $L$ of the previous $n - 1$ input vectors are used). The novelty criterion sparsification method was used for the tests (described in Ref. [2]).

Note that the convergence behavior can vary based on the nonlinear channel, as well as the kernel used for learning. Thus, the number of iterations to converge may vary. To illustrate the complexity/performance trade off, here we provide simulation results with the following settings.

The channel used was (based on the Wiener nonlinear model):

$$x(n) = (-0.9 + 0.8j)u(n) + (0.6 - 0.7j)u(n-1) \quad \text{(linear portion)}$$

$$y(n) = x(n) - (0.1 + 0.15j)x^2(n) + (0.06 + 0.05j)x^3(n) + v(n)$$
$$\text{(nonlinearity + noise)}$$

where $u(n)$ has unit variance, and $v(n)$ has variance 0.006 (−22 dBm).

And the kernel functions used.

Complex polynomial kernel:

$$\kappa(\mathbf{u}, \mathbf{v}) = \left(1 + \mathbf{u}^H \mathbf{v}\right)^q, \quad \text{with } q = 4$$

Complex Gaussian kernel:

$$\kappa(\mathbf{u}, \mathbf{v}) = \exp\left(-\sum_{i=1}^{M} (u_i - v_i^*)^2/\sigma^2\right), \quad \text{with } \sigma = 3.16$$

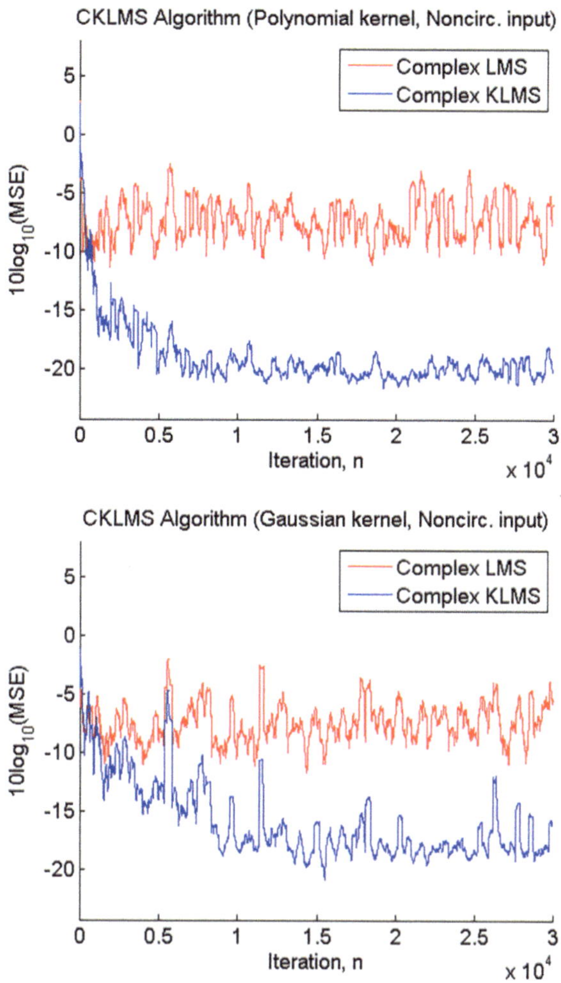


**Fig. B.1** Complex KLMS using polynomial, Gaussian kernel

(where **u**, **v** are of length $M = 5$).

Results for channel estimation are shown in Fig. B.1 (for 30,000 iterations).

For the channel estimation task here, both kernel functions provided about 8–10 dB of performance benefit. The computational complexity in each case was (note $M = 5$ was used for all cases).

For Complex LMS:

$$\begin{aligned}\text{Operations/iter.} &= 2M + 1 \text{ mult's, } \ 2M \text{ adds}\\ &= 11 \text{ mult's, } \ 10 \text{ adds}\end{aligned}$$

For Complex KLMS (Polynomial): $L = 1390$

$$\begin{aligned}\text{Operations/iter.} &= (M+1)L+1 \text{ mult's},\ (M+1)L \text{ adds}, L \text{ exponent eval's}\\ &= 6L+1 \text{ mult's},\ 6L \text{ adds}, L \text{ exponent eval's}\\ &= 8341 \text{ mult's},\ 8340 \text{ adds},\ 1390 \text{ exponent eval's}\end{aligned}$$

For Complex KLMS (Gaussian): $L = 4178$

$$\begin{aligned}\text{Operations/iter.} &= (M+2)L+1 \text{ mult's}, 2ML \text{ adds},\ L \exp(.) \text{ eval's}\\ &= 7L+1 \text{ mult's},\ 10L \text{ adds},\ L \exp(.) \text{ eval's}\\ &= 29{,}247 \text{ mult's},\ 41{,}780 \text{ adds},\ 4178 \exp(.) \text{ eval's}\end{aligned}$$

From the results, we can see that careful consideration of the transform for the application (in this case, use of a polynomial kernel) can provide both performance benefit and reduced complexity. Though the Gaussian kernel is suitable for universal function approximation, specific knowledge of the structure of a channel may lead to an alternative choice of kernel function for improved performance (i.e. a transform space with reduced dimension size, and thus faster convergence).

For applications where an increase in complexity is justified by the performance gain (which may depend on the platform being considered), the Complex Kernel LMS may be a good approach. In particular, the approach allows learning on an infinite-dimensional space with finite complexity [2], which makes it ideal for adaptive learning in scenarios where the nonlinearity of the channel is unknown and substantial.

# Appendix C
# Convergence of CKLMS: Finding $\mathbf{R}_{\kappa\kappa}$, Complexified Kernel

The following describes the results for evaluating the correlation matrix $\mathbf{R}_{\kappa\kappa}$ based on the complexified Gaussian kernel (i.e. Eq. (9.7) from Chap. 9).

For evaluating the elements of matrix $\mathbf{R}_{\kappa\kappa}$, i.e.

$$[\mathbf{R}_{\kappa\kappa}]_{ij} = \begin{cases} E\big[|\kappa(\mathbf{u}(\omega_i), \mathbf{u}(n))|^2\big], & i = j \\ E\big[\kappa(\mathbf{u}(\omega_i), \mathbf{u}(n)) \cdot \kappa^*(\mathbf{u}(\omega_j), \mathbf{u}(n))\big], & i \neq j \end{cases} \tag{C.1}$$

where $1 \leq i, j \leq M$, the results for the complexified Gaussian kernel are shown here.

(1) For the $i = j$ case:

$$E\big[|\kappa(\mathbf{u}(\omega_i), \mathbf{u}(n))|^2\big] = E\left[\exp\left[\frac{-2}{2\xi^2}\left(\sum_{k=1}^{L} \mathrm{Re}\{u_k(n) - u_k(\omega_i)\}^2 + \sum_{k=1}^{L} \mathrm{Im}\{u_k(n) - u_k(\omega_i)\}^2\right)\right]\right] \tag{C.2}$$

Considering the input signal model (9.2) of Chap. 9, we can form the vector $\mathbf{y} = [\mathbf{x}_{re}{}^T(n)\ \mathbf{x}_{im}{}^T(n)\ \mathbf{x}_{re}{}^T(\omega_i)\ \mathbf{x}_{im}{}^T(\omega_i)]^T$, and write

$$\begin{aligned} &2\sum_{k=1}^{L} \mathrm{Re}\{u_k(n) - u_k(\omega_i)\}^2 + 2\sum_{k=1}^{L} \mathrm{Im}\{u_k(n) - u_k(\omega_i)\}^2 \\ &= 2\sum_{k=1}^{L} \left[\rho_{re} x_{re,k}(n) - \rho_{re} x_{re,k}(\omega_i)\right]^2 + 2\sum_{k=1}^{L} \left[\rho_{im} x_{im,k}(n) - \rho_{im} x_{im,k}(\omega_i)\right]^2 \\ &= \mathbf{y}^T \mathbf{O}_{r1} \mathbf{y} \end{aligned}$$

where



T. K. Paul and T. Ogunfunmi, *Adaptive Filter Algorithms For Complex and Quaternion Data*, https://doi.org/10.1007/978-3-032-23798-9

$$\mathbf{O}_{r1} = \begin{bmatrix} 2\rho_{re}^2\mathbf{I} & 0 & -2\rho_{re}^2\mathbf{I} & 0 \\ 0 & 2\rho_{im}^2\mathbf{I} & 0 & -2\rho_{im}^2\mathbf{I} \\ -2\rho_{re}^2\mathbf{I} & 0 & 2\rho_{re}^2\mathbf{I} & 0 \\ 0 & -2\rho_{im}^2\mathbf{I} & 0 & 2\rho_{im}^2\mathbf{I} \end{bmatrix} \tag{C.3}$$

and $\mathbf{I}$ is the $L \times L$ identity matrix.

For convenience, matrix $\mathbf{O}_{r1}$ can be expressed as

$$\mathbf{O}_{r1} = \begin{bmatrix} 2 & -2 \\ -2 & 2 \end{bmatrix} \otimes \begin{bmatrix} \rho_{re}^2\mathbf{I} & 0 \\ 0 & \rho_{im}^2\mathbf{I} \end{bmatrix} \tag{C.4}$$

where $\otimes$ is the Kronecker product.

Thus (C.2) can be written as

$$E\left[|\kappa(\mathbf{u}(\omega_i), \mathbf{u}(n))|^2\right] = E\left[\exp\left(\frac{-1}{2\xi^2}\mathbf{y}^T\mathbf{O}_{r1}\mathbf{y}\right)\right] \tag{C.5}$$

Then, using the characteristic function (9.30), we get

$$E\{e^{-(\mathbf{y}^T\mathbf{O}_{r1}\mathbf{y})/(2\xi^2)}\} = \det(\mathbf{I} + \mathbf{O}_{r1}\mathbf{R}_{\mathbf{y}}/\xi^2)^{-1/2} \tag{C.6}$$

with matrix $\mathbf{O}_{r1}$, vector $\mathbf{y}$, described previously.

(2) For the $i \neq j$ case:

$$\begin{aligned} &E\left[\kappa(\mathbf{u}(\omega_i), \mathbf{u}(n)) \cdot \kappa^*(\mathbf{u}(\omega_j), \mathbf{u}(n))\right] \\ &\quad = E\left[\exp\left[\frac{-1}{2\xi^2}\left(\sum_{k=1}^{L}\mathrm{Re}\{u_k(n) - u_k(\omega_i)\}^2 + \sum_{k=1}^{L}\mathrm{Im}\{u_k(n) - u_k(\omega_i)\}^2\right.\right.\right. \\ &\qquad \left.\left.\left. + \sum_{k=1}^{L}\mathrm{Re}\left\{u_k(n) - u_k(\omega_j)\right\}^2 + \sum_{k=1}^{L}\mathrm{Im}\left\{u_k(n) - u_k(\omega_j)\right\}^2\right)\right]\right] \end{aligned} \tag{C.7}$$

Forming vector $\mathbf{y} = [\mathbf{x}_{re}{}^T(n)\ \mathbf{x}_{im}{}^T(n)\ \mathbf{x}_{re}{}^T(\omega_i)\ \mathbf{x}_{im}{}^T(\omega_i)\ \mathbf{x}_{re}{}^T(\omega_j)\ \mathbf{x}_{im}{}^T(\omega_j)]^T$, we can write

$$\begin{aligned} &\sum_{k=1}^{L}\mathrm{Re}\{u_k(n) - u_k(\omega_i)\}^2 + \sum_{k=1}^{L}\mathrm{Im}\{u_k(n) - u_k(\omega_i)\}^2 \\ &\quad + \sum_{k=1}^{L}\mathrm{Re}\left\{u_k(n) - u_k(\omega_j)\right\}^2 + \sum_{k=1}^{L}\mathrm{Im}\left\{u_k(n) - u_k(\omega_j)\right\}^2 = \mathbf{y}^T\mathbf{O}_{r2}\mathbf{y} \end{aligned} \tag{C.8}$$

where

$$\mathbf{O}_{r2} = \begin{bmatrix} 2 & -1 & -1 \\ -1 & 1 & 0 \\ -1 & 0 & 1 \end{bmatrix} \otimes \begin{bmatrix} \rho_{re}^2 \mathbf{I} & 0 \\ 0 & \rho_{im}^2 \mathbf{I} \end{bmatrix} \tag{C.9}$$

Finally, based on Eq. (9.30), we have

$$E\{e^{-(\mathbf{y}^T \mathbf{O}_{r2} \mathbf{y})/(2\xi^2)}\} = \det(\mathbf{I} + \mathbf{O}_{r2}\mathbf{R}_{\mathbf{y}}/\xi^2)^{-1/2} \tag{C.10}$$

Thus (C.6), (C.10) are the elements of matrix $\mathbf{R}_{\kappa\kappa}$ of (C.1) (i.e. the $i = j$ and $i \neq j$ cases).

# Appendix D
# CKLMS Stability/Steady-State Analysis, Complexified Kernel

The stability, steady-state analysis of CKLMS using the complexified Gaussian kernel is described here. The analysis directly follows [4], given the equivalence of the $\mathbf{C}_{\mathrm{v}}(n)$ expressions (9.75) and (9.76) for CKLMS with the expressions of KLMS (although $\mu_1$ to $\mu_5$ evaluate differently).

For the stability analysis of $\eta$ for the algorithm, expression (9.40) is rewritten as

$$\mathbf{c}_{\mathrm{v}}(n+1) = \mathbf{G}\,\mathbf{c}_{\mathrm{v}}(n) + \eta^2 J_{min}\mathbf{r}_{\kappa\kappa} \tag{D.1}$$

where $\mathbf{c}_{\mathrm{v}}(n)$ is the lexicographic representation of $\mathbf{C}_{\mathrm{v}}(n)$, i.e. the columns of matrix $\mathbf{C}_{\mathrm{v}}(n)$ stacked column-wise as a vector. Vector $\mathbf{r}_{\kappa\kappa}$ is similar (i.e. lexigraphical representation of $\mathbf{R}_{\kappa\kappa}$). The $M^2 \times M^2$ matrix $\mathbf{G}$ is expressed, based on (9.75) and (9.76) for the complexified Gaussian kernel:

$$\mathbf{G} = [\mathbf{h}^{11} \;\; \mathbf{h}^{12} \;\; \cdots \;\; \mathbf{h}^{1M} \;\; \cdots \;\; \mathbf{h}^{MM}] \tag{D.2}$$

where $\mathbf{h}^{ij}$ is the $M^2 \times 1$ lexicographic representation of the $M \times M$ matrix $\mathbf{H}^{ij}$, whose elements are defined as

$$\text{if } (i=j): \begin{cases} [\mathbf{H}^{ii}]_{ii} = 1 - 2\eta r_{md} + \eta^2\mu_1, & \\ \mathbf{H}^{ii}]_{pp} = \eta^2\mu_3, & p \neq i \\ [\mathbf{H}^{ii}]_{ip} = \eta^2\mu_2 - \eta r_{od} = [\mathbf{H}^{ii}]_{pi}, & p \neq i \\ [\mathbf{H}^{ii}]_{pl} = \eta^2\mu_4, & \text{otherwise} \end{cases} \tag{D.3}$$

$$\text{if } (i \neq j): \begin{cases} [\mathbf{H}^{ij}]_{ij} = [\mathbf{H}^{ij}]_{ji} = \frac{1}{2}(1 - 2\eta r_{md} + 2\eta^2\mu_3) & \\ [\mathbf{H}^{ij}]_{pp} = \eta^2\mu_4, & p \neq i,j \\ [\mathbf{H}^{ij}]_{ii} = [\mathbf{H}^{ij}]_{jj} = \eta^2\mu_2 - \eta r_{od}, & \\ [\mathbf{H}^{ij}]_{ip} = [\mathbf{H}^{ij}]_{pi} = \frac{1}{2}(2\eta^2\mu_4 - \eta r_{od}) & p \neq i,j \\ [\mathbf{H}^{ij}]_{pj} = [\mathbf{H}^{ij}]_{jp} = \frac{1}{2}(2\eta^2\mu_4 - \eta r_{od}) & p \neq i,j \\ [\mathbf{H}^{ij}]_{pl} = \eta^2\mu_5, & \text{otherwise} \end{cases} \tag{D.4}$$



T. K. Paul and T. Ogunfunmi, *Adaptive Filter Algorithms For Complex and Quaternion Data*, https://doi.org/10.1007/978-3-032-23798-9

Matrix **G** can be seen to be symmetric.

Using this form for the weight-error covariance, the stability bound for $\eta$ can be found as follows. For covariance $\mathbf{c}_\mathbf{v}(n)$ to converge, a necessary and sufficient condition is that all eigenvalues of **G** be inside the open interval $(-1, 1)$. Further, it is known that all eigenvalues of **G** lie inside the union of Gerschgorin disks, where each disk is centered at a diagonal element of **G**, with radius determined by the sum of the absolute values of the remaining elements of the same row. For the stability of $\mathbf{c}_\mathbf{v}(n)$, a sufficient condition is

$$|[\mathbf{G}]_{ii}| + \sum_{\substack{l=1 \\ l \neq i}}^{M^2} |[\mathbf{G}]_{il}| < 1, \quad \text{for } i = 1, \ldots, M^2 \tag{D.5}$$

This results in two equalities when $M \geq 3$:

$$\begin{aligned}(1 - 2\eta r_{md} + \eta^2\mu_1) + (M-1)\eta^2\mu_3 \\ + 2(M-1)|\eta^2\mu_2 - \eta r_{od}| + (M-1)(M-2)\eta^2\mu_4 < 1,\end{aligned} \tag{D.6}$$

$$\begin{aligned}(1 - 2\eta r_{md} + 2\eta^2\mu_3) + 2|\eta^2\mu_2 - \eta r_{od}| + (M-2)\eta^2\mu_4 \\ + 2(M-2)|2\eta^2\mu_4 - \eta r_{od}| + (M-2)(M-3)\eta^2\mu_5 < 1\end{aligned} \tag{D.7}$$

The $\eta$ stability bound for KLMS (applicable for CKLMS with a complexified Gaussian kernel) is summarized here.

For considering (D.6), we initially compute [4]:

$$\eta_1 := \frac{2r_{md} + 2(M-1)r_{od}}{\mu_1 + 2(M-1)\mu_2 + (M-1)\mu_3 + (M-1)(M-2)\mu_4}, \qquad \eta_2 := \frac{\theta_1}{\theta_2} \tag{D.8}$$

$$\begin{aligned}\theta_1 &= 2r_{md} - 2(M-1)r_{od}, \\ \theta_2 &= \mu_1 - 2(M-1)\mu_2 + (M-1)\mu_3 + (M-1)(M-2)\mu_4\end{aligned}$$

Then, a constraint for $\eta$ can be computed with the following pseudo-code:

$$\text{If } (\theta_1 > 0 \text{ and } \theta_2 > 0): \quad \eta < \min\{\eta_1, \eta_2\} \tag{D.9}$$

If $(\theta_1 > 0$ and $\theta_2 < 0)$ or $(\theta_1 < 0$ and $\theta_2 < 0)$ : $\eta < \eta_1$
If $(\theta_1 < 0$ and $\theta_2 > 0)$ : $\eta$ found from eigenvalues of **G**

For the final case (i.e. $\theta_1 < 0$ and $\theta_2 > 0$), the eigenvalues of **G** are evaluated numerically and used to determine $\eta$. This case should be tested right away to avoid added computation.

Considering (D.7), additional constraints for step-size $\eta$ are

$$\eta_3 := \frac{2(M-2)(r_{md}+(M-1)r_{od})}{\theta_3}, \qquad \eta_4 := (M-2)\frac{\theta_1}{\theta_4} \tag{D.10}$$

$$\theta_3 = -\mu_1 + (M-1)\mu_3 + 4(M-1)(M-2)\mu_4 + (M-1)(M-2)(M-3)\mu_5$$

$$\theta_4 = -\mu_1 + (M-1)\mu_3 - 4(M-1)(M-2)\mu_4 + (M-1)(M-2)(M-3)\mu_5$$

The added constraints, applied to the output of (D.9), are

$$\begin{array}{ll}
\text{If } \theta_3 > 0 \text{ then :} & \\
\text{If } (\theta_1 > 0 \text{ and } \theta_4 > 0): & \eta < \min\{\eta_3, \eta_4\} \\
\text{If } (\theta_1 > 0 \text{ and } \theta_4 < 0) \text{ or } (\theta_1 < 0 \text{ and } \theta_4 < 0): & \eta < \eta_3 \\
\text{If } (\theta_1 < 0 \text{ and } \theta_4 > 0): & \eta \text{ found from } \mathbf{G} \text{ eigen values} \\
\text{If } \theta_3 < 0 \text{ then :} & \\
\text{If } (\theta_1 > 0 \text{ and } \theta_4 > 0): & \eta < \eta_4 \\
\text{If } (\theta_1 > 0 \text{ and } \theta_4 < 0) \text{ or } (\theta_1 < 0 \text{ and } \theta_4 < 0): & \eta > 0 \\
\text{If } (\theta_1 < 0 \text{ and } \theta_4 > 0): & \eta \text{ found from } \mathbf{G} \text{ eigen values}
\end{array} \tag{D.11}$$

Thus, we have described an approach for computing sufficient stability conditions for $\eta$ (using Gerschgorin disks).

The steady state behavior analysis of [4] can also be applied to the complexified Gaussian kernel, and we summarize the analysis here. For the steady-state value of $\mathbf{c}_v(n)$, writing the closed-form solution of $\mathbf{c}_v(n)$ as

$$\mathbf{c}_v(n) = \mathbf{G}^n[\mathbf{c}_v(0) - \mathbf{c}_v(\infty)] + \mathbf{c}_v(\infty) \tag{D.12}$$

The vector at steady-state is

$$\mathbf{c}_v(\infty) = \eta^2 J_{min}(\mathbf{I} - \mathbf{G})^{-1}\mathbf{r}_{\kappa\kappa} \tag{D.13}$$

The steady-state solution for $\mathbf{c}_v(n)$ can be found based on $\mathbf{C}_v(n)$. From (9.40), for matrix $\mathbf{C}_v(n)$ in steady state:

$$\mathbf{R}_{\kappa\kappa}\mathbf{C}_v(\infty) + \mathbf{C}_v(\infty)\mathbf{R}_{\kappa\kappa} - \eta\mathbf{T}(n) = \eta\mathbf{R}_{\kappa\kappa}J_{min} \tag{D.14}$$

From [4], $\mathbf{C}_v(\infty)$ has the same form as $\mathbf{R}_{\kappa\kappa}$, i.e. all the main diagonal elements are equal, and all the off-diagonal elements are equal. In particular, these elements were shown to be

$$c_{md} = \frac{\eta J_{\min}}{2r_{md} - \eta t_1} \times (r_{md} +$$

$$\frac{((2r_{md} - \eta t_1)r_{od} - (2r_{od} - \eta t_3)r_{md})(\eta t_2 - 2(M-1)r_{od})}{(\eta t_2 - 2(M-1)r_{od})(2r_{od} - \eta t_3) - (\eta t_4 - 2(r_{md} + (M-2)r_{od}))(2r_{md} - \eta t_1)}) \tag{D.15}$$

$$c_{od} = \frac{\eta J_{\min}((2r_{md} - \eta t_1)r_{od} - (2r_{od} - \eta t_3)r_{md})}{(\eta t_2 - 2(M-1)r_{od})(2r_{od} - \eta t_3) - (\eta t_4 - 2(r_{md} + (M-2)r_{od}))(2r_{md} - \eta t_1)} \tag{D.16}$$

where

$$t_1 = \mu_1 + (M-1)\mu_3, \qquad t_2 = 2(M-1)\mu_2 + (M-1)(M-2)\mu_4$$

$$t_3 = 2\mu_2 + (M-2)\mu_4, \qquad t_4 = 4(M-2)\mu_4 + 2\mu_3 + (M-3)(M-2)\mu_5$$

Thus, using (9.23), the steady-state MSE can be seen to be

$$J_{ms}(\infty) = J_{min} + M(r_{md}c_{md} + (M-1)r_{od}c_{od}) \tag{D.17}$$

Design guidelines for CKLMS parameter selection based on the convergence analysis are described here. Sparsification rule (9.12) is assumed for designing the dictionary, however any technique for sparsification may be considered.

With the design goal of achieving an MSE less than some specified $J_{max}$, the procedure is

1. Select a coherence threshold $\varepsilon_0$, and range of kernel bandwidths $\xi$ to test. Parameter $\varepsilon_0$ is selected so that the resulting value of $M$ is not excessive for each $\xi$ value (for reduced computation). The range of $\xi$ values is chosen for meeting the specified $J_{max}$. For each $(\xi, \varepsilon_0)$, $M$ is found by training the dictionary using input $\mathbf{u}(n)$ until the dictionary size stabilizes. The process is repeated for numerous realizations, and the average dictionary size used to determine $M$.
2. Determine the desired output $d(n)$, based on input $\mathbf{u}(n)$, and estimate $E\{d^2(n)\}$ and $\mathbf{p}_{\mathrm{kd}}$ for several runs. Note that input $\mathbf{u}(n)$ is affected by the choice of circularity parameter $\rho$, and $d(n)$ is typically the output of an unknown system.
3. Find $J_{min}$ using (9.17) for each set $(M, \xi)$ pair. If $J_{max} > J_{min}$ does not occur for any pair, then go back to step 1, choosing different $\varepsilon_0$, $\xi$ values.
4. Compute $\eta_{max}$ based on (D.9) and (D.11) for each $\xi$, and select the largest $\eta$ so that $\eta < \eta_{max}$ and $J_{ms}(\infty) < J_{max}$, where $J_{ms}(\infty)$ is computed with (D.17).
5. Using $M$, $\mathbf{R}_{\mathrm{KK}}$, $\mathbf{p}_{\mathrm{kd}}$, $J_{min}$ and $\eta$, find $\mathbf{G}$ using (D.2), and $\mathbf{c}_{\mathrm{v}}(n)$ from (D.1).
6. Compute $\mathbf{C}_{\mathrm{v}}(\infty)$ using (D.15), (D.16), then $J_{ms}(\infty)$ using (D.17).
7. Choose design parameter $\epsilon$ for the convergence criterion: $\|\mathbf{c}_{\mathrm{v}}(n)\text{-}\mathbf{c}_{\mathrm{v}}(n_\epsilon)\| < \epsilon$, and find $n_\epsilon$ from simulations based on the parameters from previous steps. Repeat steps 1–7 for all $\xi$ in the set, and choose the one that yields the minimum $n_\epsilon$.

The rules described allow both the MSE and convergence rate performance to be considered, although they may exhibit varying convergence rates depending on circularity $\rho$.

# Appendix E
# Gaussian Kernel for Quaternion Data

Here, we continue to analyze the proposed quaternion Gaussian kernel:

$$\kappa_{\mathrm{Q}}(\mathbf{a}, \mathbf{b}) = \exp\left(-\sigma^2 \sum_{i=1}^{d} (a_i - b_i^*)^2\right) \tag{E.1}$$

based on the kernel form (10.12b) from Sect. 10.2 (repeated here):

$$\kappa_{\mathrm{Q}_{IB}}(\mathbf{a}, \mathbf{b}) = \exp\left(-\sigma^2 \sum_{i=1}^{d} 4\mathrm{Re}\{(a_i - b_i^*)^2\}\right) \tag{E.2}$$

Equation (E.2), in the $d = 1$ case, eexpanded, and only keeping terms related to the inner product between **a** and **b**, leads to

$$\sum_{n=0}^{\infty} \frac{\left(4\sigma^2\mathrm{Re}\{ab^* + b^*a\}\right)^n}{n!}$$
$$= \sum_{n=0}^{\infty} \frac{\left(\sigma^2[ab^* + a^i b^{*i} + a^j b^{*j} + a^k b^{*k} + a^* b + a^{*i} b^i + a^{*j} b^j + a^{**k} b^k]\right)^n}{n!} \tag{E.3}$$

The $n = 1$ term was shown to be

$$\sigma^2(a^* b + a^{*i} b^i + a^{*j} b^j + a^{*k} b^k + ab^* + a^i b^{*i} + a^j b^{*j} + a^k b^{*k})$$
$$= \boldsymbol{\psi}_1^H(a)\boldsymbol{\psi}_1(b)$$

where $\psi_1(x) = \sigma[x, x^i, x^j, x^k, x^*, x^{*i}, x^{*j}, x^{*k}]^T$. Here we also consider the $n > 1$ terms.

For the $n = 2$ term:

$$(\sigma^4/2!)(a^* b + a^{*i} b^i + a^{*j} b^j + a^{*k} b^k + ab^* + a^i b^{*i} + a^j b^{*j} + a^k b^{*k})^2$$



T. K. Paul and T. Ogunfunmi, *Adaptive Filter Algorithms For Complex and Quaternion Data*, https://doi.org/10.1007/978-3-032-23798-9

$$
\begin{aligned}
&= (\sigma^4/2!)[(a^*ba^*b + a^*ba^{*i}b^i + a^*ba^{*j}b^j + \ldots + a^*ba^kb^{*k}) \\
&\quad + (a^{*i}b^ia^*b + a^{*i}b^ia^{*i}b^i + a^{*i}b^ia^{*j}b^j + \ldots + a^{*i}b^ia^kb^{*k}) \\
&\quad\ \vdots \\
&\quad + (a^kb^{*k}a^*b + a^kb^{*k}a^{*i}b^i + a^kb^{*k}a^{*j}b^j + \ldots + a^kb^{*k}a^kb^{*k})]
\end{aligned}
\tag{E.4}
$$

Thus, for $n = 2$, the term is not readily expressible as an inner product considering non-commutative quaternion multiplication.

However, we consider factoring in the following way. For evaluating term $a^*ba^*b$ from (E.4):

$$
\begin{aligned}
a^*ba^*b &= \tfrac{1}{2}a^*(a^ib^{*i} + a^jb^{*j} + a^kb^{*k} - ab^*)b \\
&= \tfrac{1}{2}(a^*a^ib^{*i}b + a^*a^jb^{*j}b + a^*a^kb^{*k}b - a^*ab^*b)
\end{aligned}
\tag{E.5}
$$

where the conjugate of $ba^*$ is evaluated based on the quaternion properties of Chap. 5. Table E.1 shows the first half of all terms of (E.4) similarly expanded, and their equivalent sum. The remaining half can be similarly shown.

From this table, the $n = 2$ case may also be expressed using an inner product form, yielding

$$
\begin{aligned}
&(\sigma^4/2!)(a^*b + a^{*i}b^i + a^{*j}b^j + a^{*k}b^k + ab^* + a^ib^{*i} + a^jb^{*j} + a^kb^{*k})^2 \\
&= \boldsymbol{\psi}_2^H(a)\boldsymbol{\psi}_2(b)
\end{aligned}
$$

where $\psi_2(x) = (\sigma^4/2!)^{1/2}[xx,\ xx^i,\ xx^j,\ xx^k,\ x^ix,\ x^ix^i,\ \ldots,\ x^{*k}x^{*k}]^T$. The vector $\psi_2(x)$ consists of 64 elements, which are all the permutations of the multiplication of two elements chosen, with repetition, from the set: $\{x, x^i, x^j, x^k, x^*, x^{*i}, x^{*j}, x^{*k}\}$.

A similar process for larger $n$ can be shown to yield

$$
\begin{aligned}
&(\sigma^{2n}/n!)(a^*b + a^{*i}b^i + a^{*j}b^j + a^{*k}b^k + ab^* + a^ib^{*i} + a^jb^{*j} + a^kb^{*k})^n \\
&= \boldsymbol{\psi}_n^H(a)\boldsymbol{\psi}_n(b)
\end{aligned}
\tag{E.6}
$$

where $\psi_n(x) = (\sigma^{2n}/n!)^{1/2}[x^{\cdots}xx,\ x^{\cdots}xx^i,\ x^{\cdots}xx^j,\ \ldots,\ x^{*k\cdots}x^{*k}x^{*k}]^T$, and consists of $8^n$ elements, which are all the permutations of the multiplication of $n$ elements chosen, with repetition, from the set: $\{x, x^i, x^j, x^k, x^*, x^{*i}, x^{*j}, x^{*k}\}$.

Finally, combining with Eq. (10.14) yields

$$
\begin{aligned}
\kappa_{Q_{IB}}(a, b) &= \exp\left(-\sigma^2 4\mathrm{Re}\{(a - b^*)^2\}\right) \\
&= \boldsymbol{\varphi}^H(a)\boldsymbol{\varphi}(b)
\end{aligned}
\tag{E.7}
$$

where $\boldsymbol{\varphi}(x) = \exp(-4\sigma^2\mathrm{Re}\{x^2\})[\boldsymbol{\psi}_0^T(x),\ \boldsymbol{\psi}_1^T(x),\ \boldsymbol{\psi}_2^T(x),\ \boldsymbol{\psi}_3^T(x)\ \ldots\ ]^T$.

From (E.7), we see that the Mercer condition applies to the function $\kappa_{Q_{IB}}(\mathbf{a}, \mathbf{b})$ of (E.2) when $d = 1$ (i.e. the scalar case). The vector case should have a similar factoring

**Table E.1** Expansion and Factoring of Equation (E.4)

| | |
|---|---|
| $a^*ba^*b = \frac{1}{2}\left[a^*a^ib^{*i}b + a^*a^jb^{*j}b + a^*a^kb^{*k}b - a^*ab^*b\right]$ | sum $= a^*ab^*b$ |
| $a^*ba^{*i}b^i = \frac{1}{2}\left[a^*ab^{*i}b^i + a^*a^kb^{*j}b^i + a^*a^jb^{*k}b^i - a^*a^ib^*b^i\right]$ | $+a^*a^ib^{*i}b$ |
| $a^*ba^{*j}b^j = \frac{1}{2}\left[a^*a^kb^{*i}b^j + a^*ab^{*j}b^j + a^*a^ib^{*k}b^j - a^*a^jb^*b^j\right]$ | $+a^*a^jb^{*j}b$ |
| $a^*ba^{*k}b^k = \frac{1}{2}\left[a^*a^jb^{*i}b^k + a^*a^ib^{*j}b^k + a^*ab^{*k}b^k - a^*a^kb^*b^k\right]$ | $+a^*a^kb^{*k}b$ |
| $a^*bab^* = \frac{1}{2}\left[a^*a^{*i}b^{*i}b^* + a^*a^{*j}b^{*j}b^* + a^*a^{*k}b^{*k}b^* - a^*a^*b^*b^*\right]$ | $+a^*a^*bb$ |
| $a^*ba^ib^{*i} = \frac{1}{2}\left[a^*a^*b^{*i}b^{*i} + a^*a^{*k}b^{*j}b^{*i} + a^*a^{*j}b^{*k}b^{*i} - a^*a^{*i}b^*b^{*i}\right]$ | $+a^*a^*bb$ |
| $a^*ba^jb^{*j} = \frac{1}{2}\left[a^*a^{*k}b^{*i}b^{*j} + a^*a^*b^{*j}b^{*j} + a^*a^{*i}b^{*k}b^{*j} - a^*a^{*j}b^*b^{*j}\right]$ | $+a^*a^*b^ib$ |
| $a^*ba^kb^{*k} = \frac{1}{2}\left[a^*a^{*j}b^{*i}b^{*k} + a^*a^{*i}b^{*j}b^{*k} + a^*a^*b^{*k}b^{*k} - a^*a^{*k}b^*b^{*k}\right]$ | $+a^*a^{**}b^jb$ |
| $a^{*i}b^ia^*b = \frac{1}{2}\left[a^{*i}a^ib^*b + a^{*i}a^jb^{*k}b + a^{*i}a^{*k}b^{*j}b - a^{*i}ab^{*i}b\right]$ | $+a^{*i}ab^*b^i$ |
| $a^{*i}b^ia^{*i}b^i = \frac{1}{2}\left[a^{*i}ab^*b^i + a^{*i}a^{*k}b^{*k}b^i + a^{*i}a^jb^{*j}b^i - a^{*i}a^ib^{*i}b^i\right]$ | $+a^{*i}a^ib^{*i}b^i$ |
| $a^{*i}b^ia^{*j}b^j = \frac{1}{2}\left[a^{*i}a^{*k}b^*b^j + a^{*i}ab^{*k}b^j + a^{*i}a^ib^{*j}b^j - a^{*i}a^jb^{*i}b^j\right]$ | $+a^{*i}a^jb^{*j}b^i$ |
| $a^{*i}b^ia^{*k}b^k = \frac{1}{2}\left[a^{*i}a^jb^*b^k + a^{*i}a^ib^{*k}b^k + a^{*i}ab^{*j}b^k - a^{*i}a^kb^{*i}b^{*k}\right]$ | $+a^{*i}a^kb^{*kk}b^i$ |
| $a^{*i}b^iab^* = \frac{1}{2}\left[a^{*i}a^{*i}b^*b^* + a^{*i}a^{*j}b^{*k}b^* + a^{*i}a^{*k}b^{*j}b^* - a^{*i}a^*b^{*i}b^*\right]$ | $+a^{*i}a^*bb^i$ |
| $a^{*i}b^ia^ib^{*i} = \frac{1}{2}\left[a^{*i}a^*b^*b^{*i} + a^{*i}a^{*k}b^{*k}b^{*i} + a^{*i}a^{*j}b^{*j}b^{*i} - a^{*i}a^{*i}b^{*i}b^{*i}\right]$ | $+a^{*i}a^{*i}b^ib^i$ |
| $a^{*i}b^ia^jb^{*j} = \frac{1}{2}\left[a^{*i}a^{*k}b^*b^{*j} + a^{*i}a^*b^{*k}b^{*j} + a^{*i}a^{*i}b^{*j}b^{*j} - a^{*i}a^{*j}b^{*i}b^{*j}\right]$ | $+a^{*}_{i}a^{*j}b^jb^i$ |
| $a^{*i}b^ia^kb^{*k} = \frac{1}{2}\left[a^{*i}a^{*j}b^*b^{*k} + a^{*i}a^{*i}b^{*k}b^{*k} + a^{*i}a^*b^{*j}b^{*k} - a^{*i}a^{*k}b^{*i}b^{*k}\right]$ | $+a^{*i}a^{*k^k}b^kb^i$ |
| $a^{*j}b^ja^*b = \frac{1}{2}\left[a^{*j}a^ib^{*k}b + a^{*j}a^jb^*b + a^{*j}a^kb^{*i}b - a^{*j}ab^{*j}b\right]$ | $+a^{*j}ab^*b^j$ |
| $a^{*j}b^ja^{*i}b^i = \frac{1}{2}\left[a^{*j}ab^{*k}b^i + a^{*j}a^kb^*b^i + a^{*j}a^jb^{*i}b^i - a^{*j}a^ib^{*j}b^i\right]$ | $+a^{*j}a^ib^{*i}b^j$ |
| $a^{*j}b^ja^{*j}b^j = \frac{1}{2}\left[a^{*j}a^kb^{*k}b^j + a^{*j}ab^*b^j + a^{*j}a^ib^{*i}b^j - a^{*j}a^jb^{*j}b^j\right]$ | $+a^{*j}a^jb^{*j}b^j$ |
| $a^{*j}b^ja^{*k}b^k = \frac{1}{2}\left[a^{*j}a^jb^{*k}b^k + a^{*j}a^ib^*b^k + a^{*j}ab^{*i}b^k - a^{*j}a^kb^{*j}b^k\right]$ | $+a^{*j}a^kb^{*k}b^j$ |
| $a^{*j}b^jab^* = \frac{1}{2}\left[a^{*j}a^{*i}b^{**k}b^* + a^{*j}a^{*j}b^*b^* + a^{*j}a^{**k}b^{*i}b^* - a^{*j}a^*b^{*j}b^*\right]$ | $+a^{*j}a^*bb^j$ |
| $a^{*j}b^ja^ib^{*i} = \frac{1}{2}\left[a^{*j}a^*b^{*k}b^{*i} + a^{*j}a^{*k}b^*b^{*i} + a^{*j}a^{*j}b^{*i}b^{*i} - a^{*j}a^{*i}b^{*j}b^{*i}\right]$ | $+a^{*j}a^{*i}b^ib^j$ |
| $a^{*j}b^ja^jb^{*j} = \frac{1}{2}\left[a^{*j}a^{*k}b^{*k}b^{*j} + a^{*j}a^*b^*b^{*j} + a^{*j}a^{*i}b^{*i}b^{*j} - a^{*j}a^{*j}b^{*j}b^{*j}\right]$ | $+a^{*j}a^{*j}b^jb^j$ |
| $a^{*j}b^ja^kb^{*k} = \frac{1}{2}\left[a^{*j}a^{*j}b^{*k}b^{*k} + a^{*j}a^*b^*b^{*k} + a^{*j}a^*b^{*i}b^{*k} - a^{*j}a^{*k}b^{*j}b^{*k}\right]$ | $+a^{*j}a^{*k}b^kb^j$ |
| $a^{*k}b^ka^*b = \frac{1}{2}\left[a^{*k}a^ib^{*j}b + a^{*k}a^jb^{*i}b + a^{*k}a^kb^*b - a^{*k}ab^{*k}b\right]$ | $+a^{*kk}ab^*b^k$ |
| $a^{*k}b^ka^{*i}b^i = \frac{1}{2}\left[a^{*k}ab^{*j}b^i + a^{*k}a^kb^*b^i + a^{*k}a^jb^*b^i - a^{*k}a^ib^{*k}b^i\right]$ | $+a^{*k}a^ib^{*i}b^k$ |
| $a^{*k}b^ka^{*j}b^j = \frac{1}{2}\left[a^{*k}a^kb^{*j}b^j + a^{*k}ab^{*i}b^j + a^{*k}a^ib^*b^j - a^{*k}a^jb^{*k}b^j\right]$ | $+a^{*kk}a^jb^{*j}b^k$ |
| $a^{*kk}b^ka^{*kk}b^k = \frac{1}{2}\left[a^{*kk}a^jb^{*j}b^k + a^{*k}a^ib^{*i}b^k + a^{*kk}ab^*b^k - a^{*kk}a^kb^{*kk}b^k\right]$ | $+a^{*k}a^kb^{*k}b^k$ |
| $a^{*k}b^kab^* = \frac{1}{2}\left[a^{*kk}a^{*i}b^{*j}b^* + a^{*k}a^{*j}b^{*i}b^* + a^{*k}a^{*k}b^*b^* - a^{*k}a^*b^{*k}b^*\right]$ | $+a^{*k}a^*bb^k$ |
| $a^{*k}b^ka^ib^{*i} = \frac{1}{2}\left[a^{*k}a^*b^*b^{*i} + a^{*k}a^{*k}b^{*i}b^{*i} + a^{*k}a^{*j}b^*b^{*i} - a^{*kk}a^{*i}b^{*k}b^{*i}\right]$ | $+a^{*k}a^{*i}b^ib^k$ |
| $a^{*k}b^ka^jb^{*j} = \frac{1}{2}\left[a^{*k}a^{*k}b^{*j}b^{*j} + a^{*k}a^*b^{*i}b^{*j} + a^{*k}a^*b^*b^{*j} - a^{*k}a^{*j}b^{*k}b^{*j}\right]$ | $+a^{*k}a^{*j}b^jb^k$ |
| $a^{*k}b^ka^kb^{*k} = \frac{1}{2}\left[a^{*k}a^{*j}b^{*j}b^{*k} + a^{*k}a^{*i}b^{*i}b^{*k} + a^{*k}a^*b^*b^{*k} - a^{*k}a^{*k}b^{*k}b^{*k}\right]$ | $+a^{*k}a^{*k}b^kb^k$ |

following [1], however this remains for verification. The validity of the function $\kappa_{Q_{IB}}(\mathbf{a}, \mathbf{b})$ as a kernel function may be used for determining. Whether equation (E.1) (the quaternion Gaussian kernel definition proposed) may be seen as a valid kernel. From [5], defining a kernel $\kappa : X \times X \rightarrow \mathbb{C}$ (similar to Sect. 10.1, except defined on complex field, $\mathbb{C}$) and a function $\xi : S \rightarrow X$, we can define the composite function as $\kappa \circ \xi : S \times S \rightarrow \mathbb{C}$. Thus: $(\kappa \circ \xi)\ (\mathbf{x}_1, \mathbf{x}_2) = \kappa(\xi(\mathbf{x}_1), \xi(\mathbf{x}_2))$. Function $\xi$ represents a transformation applied to the input of kernel $\kappa$. This can be seen to reflect the relationship between equation (E.1) and $\kappa_{Q_{IB}}(\mathbf{a}, \mathbf{b})$ of (E.2).

In [5], it is observed that the composite function $\kappa \circ \xi$ is also a kernel function on $S$ for any arbitrary $\xi$. The associated Hilbert space $\mathcal{H}(\kappa \circ \xi)$ is referred to as the pull-back of $\mathcal{H}(\kappa)$ along $\xi$, and can be expressed $\mathcal{H}(\kappa \circ \xi) = \{f \circ \xi\ :\ f \in \mathcal{H}(\kappa)\}$. This property implies the validity of $\kappa_{Q_{IB}}(\mathbf{a}, \mathbf{b})$ as a kernel, given (E.1) is a valid kernel. The reverse may also be seen, given a suitable inverse map. However, since the function exp(·) differs between real and quaternion input (see (10.8)), a direct verification of the RKHS of (E.1) may be preferred, which remains to be considered.

# Appendix F
# Convergence of the Quat-KLMS and Quat-KMC Algorithms

Here, we provide a brief description of the convergence of the Quat-KLMS and Quat-KMC algorithms in terms of convergence in the mean. The convergence of the WL form of Quat-KLMS and iQuat-KLMS is also considered.

## F.1 Convergence of Quat-KLMS and WL Form

Following [6], for characterizing here the convergence of Quat-KLMS, we describe the weight-error in the following way. For Quat-KLMS, the weight update is

$$\begin{aligned}\mathbf{w}_{\mathrm{Q}}(n) &= \mathbf{w}_{\mathrm{Q}}(n-1) + \eta(\tfrac{5}{4}\widehat{\mathbf{\Phi}}_{\mathrm{Q}}(\mathbf{x}(n))e^{*}(n) - \tfrac{1}{4}\widehat{\mathbf{\Phi}}_{\mathrm{Q}}(\mathbf{x}^{i}(n))e^{i*}(n)\\ &\quad - \tfrac{1}{4}\widehat{\mathbf{\Phi}}_{\mathrm{Q}}(\mathbf{x}^{j}(n))e^{j*}(n) - \tfrac{1}{4}\widehat{\mathbf{\Phi}}_{\mathrm{Q}}(\mathbf{x}^{k}(n))e^{k*}(n))\end{aligned} \tag{F.1}$$

This can be expressed as

$$\begin{aligned}\mathbf{w}_{\mathrm{Q}}(n) &= \mathbf{w}_{\mathrm{Q}}(n-1) + \eta(\tfrac{5}{4}\widehat{\mathbf{\Phi}}_{\mathrm{Q}}(\mathbf{x}(n))e^{*}(n) - \tfrac{1}{4}[\widehat{\mathbf{\Phi}}_{\mathrm{Q}}(\mathbf{x}(n))e^{*}(n)]^{i}\\ &\quad - \tfrac{1}{4}[\widehat{\mathbf{\Phi}}_{\mathrm{Q}}(\mathbf{x}(n))e^{*}(n)]^{j} - \tfrac{1}{4}[\widehat{\mathbf{\Phi}}_{\mathrm{Q}}(\mathbf{x}(n))e^{*}(n)]^{k})\end{aligned}$$

since $\widehat{\mathbf{\Phi}}^{\eta}_{\mathrm{Q}}(\mathbf{x}(n)) = \widehat{\mathbf{\Phi}}_{\mathrm{Q}}(\mathbf{x}^{\eta}(n))$ and $a^{\eta}b^{\eta} = (ab)^{\eta}$ for $\eta \in \{i, k, j\}$.

Considering the error can be expressed as

$$\begin{aligned}e(n) &= d(n) - \mathbf{w}_{\mathrm{Q}}^{H}(n-1)\widehat{\mathbf{\Phi}}_{\mathrm{Q}}(\mathbf{x}(n))\\ &= [\mathbf{w}^{oH}(n)\widehat{\mathbf{\Phi}}_{\mathrm{Q}}(\mathbf{x}(n)) + r(n)] - \mathbf{w}_{\mathrm{Q}}^{H}(n-1)\widehat{\mathbf{\Phi}}_{\mathrm{Q}}(\mathbf{x}(n))\\ &= -\mathbf{v}^{H}(n-1)\widehat{\mathbf{\Phi}}_{\mathrm{Q}}(\mathbf{x}(n)) + r(n)\end{aligned} \tag{F.2}$$



T. K. Paul and T. Ogunfunmi, *Adaptive Filter Algorithms For Complex and Quaternion Data*, https://doi.org/10.1007/978-3-032-23798-9

where $\mathbf{v}(n) = \mathbf{w}_{\mathrm{Q}}(n) - \mathbf{w}^{o}(n)$ is the weight error definition, and $r(n)$ is the measurement noise of the nonlinear system, we can write

$$
\begin{aligned}
\mathbf{w}_{\mathrm{Q}}(n) = \mathbf{w}_{\mathrm{Q}}(n-1) &- \tfrac{5\eta}{4}[\widehat{\mathbf{\Phi}}_{\mathrm{Q}}(\mathbf{x}(n))\widehat{\mathbf{\Phi}}_{\mathrm{Q}}^{H}(\mathbf{x}(n))\mathbf{v}(n-1) - \widehat{\mathbf{\Phi}}_{\mathrm{Q}}(\mathbf{x}(n))r^{*}(n)] \\
&+ \tfrac{\eta}{4}[\widehat{\mathbf{\Phi}}_{\mathrm{Q}}(\mathbf{x}(n))\widehat{\mathbf{\Phi}}_{\mathrm{Q}}^{H}(\mathbf{x}(n))\mathbf{v}(n-1) - \widehat{\mathbf{\Phi}}_{\mathrm{Q}}(\mathbf{x}(n))r^{*}(n)]^{i} \\
&+ \tfrac{\eta}{4}[\widehat{\mathbf{\Phi}}_{\mathrm{Q}}(\mathbf{x}(n))\widehat{\mathbf{\Phi}}_{\mathrm{Q}}^{H}(\mathbf{x}(n))\mathbf{v}(n-1) - \widehat{\mathbf{\Phi}}_{\mathrm{Q}}(\mathbf{x}(n))r^{*}(n)]^{j} \\
&+ \tfrac{\eta}{4}[\widehat{\mathbf{\Phi}}_{\mathrm{Q}}(\mathbf{x}(n))\widehat{\mathbf{\Phi}}_{\mathrm{Q}}^{H}(\mathbf{x}(n))\mathbf{v}(n-1) - \widehat{\mathbf{\Phi}}_{\mathrm{Q}}(\mathbf{x}(n))r^{*}(n)]^{k}
\end{aligned} \tag{F.3}
$$

The mean value of the weight error can then be expressed as

$$
\begin{aligned}
E[\mathbf{v}(n)] = E[\mathbf{v}(n-1)] &- \tfrac{5\eta}{4}E[\widehat{\mathbf{\Phi}}_{\mathrm{Q}}(\mathbf{x}(n))\widehat{\mathbf{\Phi}}_{\mathrm{Q}}^{H}(\mathbf{x}(n))]E[\mathbf{v}(n-1)] \\
&+ \tfrac{\eta}{4}E[\widehat{\mathbf{\Phi}}_{\mathrm{Q}}(\mathbf{x}(n))\widehat{\mathbf{\Phi}}_{\mathrm{Q}}^{H}(\mathbf{x}(n))]^{i}E[\mathbf{v}(n-1)]^{i} \\
&+ \tfrac{\eta}{4}E[\widehat{\mathbf{\Phi}}_{\mathrm{Q}}(\mathbf{x}(n))\widehat{\mathbf{\Phi}}_{\mathrm{Q}}^{H}(\mathbf{x}(n))]^{j}E[\mathbf{v}(n-1)]^{j} \\
&+ \tfrac{\eta}{4}E[\widehat{\mathbf{\Phi}}_{\mathrm{Q}}(\mathbf{x}(n))\widehat{\mathbf{\Phi}}_{\mathrm{Q}}^{H}(\mathbf{x}(n))]^{k}E[\mathbf{v}(n-1)]^{k} \\
&= E[\mathbf{v}(n-1)] - \tfrac{3\eta}{2}\mathbf{C}_{\mathbf{x}}E[\mathbf{v}(n-1)] + \eta\mathrm{Re}\{\mathbf{C}_{\mathbf{x}}E[\mathbf{v}(n-1)]\}
\end{aligned} \tag{F.4}
$$

where $\mathbf{C}_{\mathbf{x}} = E[\widehat{\mathbf{\Phi}}_{\mathrm{Q}}(\mathbf{x}(n))\widehat{\mathbf{\Phi}}_{\mathrm{Q}}^{H}(\mathbf{x}(n))]$ is the covariance matrix of $\mathbf{x}(n)$ in the transform space. The (F.4) result considers the involution-based relation (5.6) in the last step, and assumes $\mathbf{x}(n)$, $\mathbf{v}(n)$ are independent, and $r(n)$ is zero-mean and independent of $\mathbf{x}(n)$.

For finding an upper bound on step-size $\eta$, we can approximate $\mathrm{Re}\{\mathbf{C}_{\mathbf{x}}E[\mathbf{v}(n-1)]\}$ as zero, maximizing $\eta$ and yielding

$$
E[\mathbf{v}(n)] \;\approx\; (1 - \eta[\tfrac{3}{2}\mathbf{C}_{\mathbf{x}}])E[\mathbf{v}(n-1)] \tag{F.5}
$$

Letting $\mathbf{R} = \frac{3}{2}\mathbf{C}_{\mathbf{x}}$, the step-size bound for convergence in the mean can be expressed [6–8]

$$
0 \;<\; \eta \;<\; (2/\lambda_{\max})
$$

where $\lambda_{\max}$ is the maximum eigenvalue of $\mathbf{R}$. Note that in the case of WL-Quat-KLMS and IB-Quat-KLMS, the same expression results, except the covariance matrix definition for $\mathbf{C}_{\mathbf{x}}$ is either $\mathbf{C}_{\mathbf{x}} = E[\widehat{\mathbf{\Phi}}_{\mathrm{Q}_{WL}}(\mathbf{x}(n))\widehat{\mathbf{\Phi}}_{\mathrm{Q}_{WL}}^{H}(\mathbf{x}(n))]$ or $\mathbf{C}_{\mathbf{x}} = E[\widehat{\mathbf{\Phi}}_{\mathrm{Q}_{IB}}(\mathbf{x}(n))\widehat{\mathbf{\Phi}}_{\mathrm{Q}_{IB}}^{H}(\mathbf{x}(n))]$, respectively.

To compare $\mathbf{C}_{\mathbf{x}}$ between these algorithms, note $\widehat{\mathbf{\Phi}}_{\mathrm{Q}_{WL}}(\cdot)$ has dimensionality 4 times that of $\widehat{\mathbf{\Phi}}_{\mathrm{Q}}(\cdot)$, while $\widehat{\mathbf{\Phi}}_{\mathrm{Q}_{IB}}(\cdot)$ has an exponentially larger dimension size. This indicates the eigenvalue spread is larger in general for the WL-based methods, and thus $\eta$ is chosen to be small based on stability concerns, as opposed to convergence rate. However, the choice of kernel parameter $\sigma$ can be used to adjust the transform map, which can reduce the eigenvalue spread, in addition to improving MSE performance.

## F.2 Convergence of IQuat-KLMS

The convergence of iQuat-KLMS can be similarly described. Based on the update:

$$\mathbf{w}_Q(n) = \mathbf{w}_Q(n-1) + \eta\left(\tfrac{3}{2}\widehat{\mathbf{\Phi}}(\mathbf{x}(n))e^*(n)\right) \tag{F.6}$$

and using the error definition of (F.2), we can write

$$\mathbf{w}_Q(n) = \mathbf{w}_Q(n-1) - \eta\tfrac{3}{2}\widehat{\mathbf{\Phi}}(\mathbf{x}(n))\widehat{\mathbf{\Phi}}(\mathbf{x}(n))^H\mathbf{v}(n-1) + \eta\tfrac{3}{2}\widehat{\mathbf{\Phi}}(\mathbf{x}(n))r^*(n) \tag{F.7}$$

Then, expressing the mean value of the weight error:

$$E[\mathbf{v}(n)] = E[\mathbf{v}(n-1)] - \eta\tfrac{3}{2}\mathbf{C}_{\mathbf{x}}E[\mathbf{v}(n-1)] \tag{F.8}$$

where $\mathbf{C}_{\mathbf{x}}$ is defined the same as before.

Comparing (F.5) and (F.8), we see the upper bound on $\eta$ is unchanged. However, the derivation of (F.5) used an additional approximation on $\mathrm{Re}\{\mathbf{C}_{\mathbf{x}}E[\mathbf{v}(n-1)]\}$. Therefore, the actual value for eigenvalue $\lambda_{\max}$ of matrix $\mathbf{R}$ for Quat-KLMS is expected to be smaller than for iQuat-KLMS. The larger eigenvalues for iQuat-KLMS imply a tighter bound on step-size $\eta$, as well as generally faster convergence for the same value of $\eta$ compared to Quat-KLMS.

## F.3 Convergence of Quat-KMC

For Quat-KMC, the weight update can be written as

$$\mathbf{w}(n) = \mathbf{w}(n-1) + \eta\exp\left[\frac{-|e(n)|^2}{2\sigma^2}\right]\left[\varphi_n e^*(n) - \tfrac{1}{2}e(n)\varphi_n^*\right] \tag{F.9}$$

where $\varphi_n = \Phi(\mathbf{x}(n))$.

Combining the exponential term $\exp(-|e(n)|^2/2\sigma^2)$ with (F.2) yields

$$\exp\left[\frac{-|e(n)|^2}{2\sigma^2}\right] = \exp\left[\frac{-\left|\mathbf{v}(n-1)^H\widehat{\mathbf{\Phi}}(\mathbf{x}(n)) - r(n)\right|^2}{2\sigma^2}\right]$$

For studying the convergence, we consider the following. During initial convergence, the weight error $\mathbf{v}(n)$ is large, particularly when $\mathbf{w}(0) = 0$ is used. In this case, we may consider approximation:

$$|\mathbf{v}(n-1)^H\widehat{\mathbf{\Phi}}(\mathbf{x}(n)) - r(n)| \approx |\mathbf{v}(n-1)^H\widehat{\mathbf{\Phi}}(\mathbf{x}(n))|$$

Note that if $\sigma$ is chosen too large, then $\exp(-|e(n)|^2/2\sigma^2) \approx 1$ always, and Quat-KMC behaves like Quat-KLMS. Meanwhile, if $\sigma$ is chosen too small then: $\exp(-|e(n)|^2/2\sigma^2) \approx 0$, and the (F.9) update is negligible. Thus, the choice of $\sigma$ should be well-considered.

We assume here that $\sigma$ is chosen so $|r(n)|^2/2\sigma^2$ is close to, but non-zero. In this case $\exp(-|r(n)|^2/2\sigma^2) \approx 1$. With $\sigma$ chosen this way, if $\mathbf{v}(n)$ is large: $\exp(-|e(n)|^2/2\sigma^2) < 1$. Consider that (F.9) can be written as

$$\begin{aligned}\mathbf{w}(n) &= \mathbf{w}(n-1) - \eta\widehat{\mathbf{\Phi}}(\mathbf{x}(n))\widehat{\mathbf{\Phi}}(\mathbf{x}(n))^H\mathbf{v}(n-1)e^{-|e(n)|^2/2\sigma^2}\\ &\quad + \eta\widehat{\mathbf{\Phi}}(\mathbf{x}(n))r^*(n)e^{-|e(n)|^2/2\sigma^2}\\ &\quad + \eta\tfrac{1}{2}\widehat{\mathbf{\Phi}}(\mathbf{x}(n))\widehat{\mathbf{\Phi}}(\mathbf{x}(n))^H\mathbf{v}(n-1)e^{-|e(n)|^2/2\sigma^2} - \eta\tfrac{1}{2}\widehat{\mathbf{\Phi}}(\mathbf{x}(n))r^*(n)e^{-|e(n)|^2/2\sigma^2}\\ &\quad - \eta\Im\{\widehat{\mathbf{\Phi}}(\mathbf{x}(n))\widehat{\mathbf{\Phi}}(\mathbf{x}(n))^H\mathbf{v}(n-1)e^{-|e(n)|^2/2\sigma^2} - \widehat{\mathbf{\Phi}}(\mathbf{x}(n))r^*(n)e^{-|e(n)|^2/2\sigma^2}\}\end{aligned} \tag{F.10}$$

and (F.1) can also be expressed:

$$\begin{aligned}\mathbf{w}_Q(n) &= \mathbf{w}_Q(n-1) - \eta\widehat{\mathbf{\Phi}}(\mathbf{x}(n))\widehat{\mathbf{\Phi}}(\mathbf{x}(n))^H\mathbf{v}(n-1) + \eta\widehat{\mathbf{\Phi}}(\mathbf{x}(n))r^*(n)\\ &\quad + \eta\tfrac{1}{2}\widehat{\mathbf{\Phi}}(\mathbf{x}(n))\widehat{\mathbf{\Phi}}(\mathbf{x}(n))^H\mathbf{v}(n-1) - \eta\tfrac{1}{2}\widehat{\mathbf{\Phi}}(\mathbf{x}(n))r^*(n)\\ &\quad - \eta\Im\{\widehat{\mathbf{\Phi}}(\mathbf{x}(n))\widehat{\mathbf{\Phi}}(\mathbf{x}(n))^H\mathbf{v}(n-1) - \widehat{\mathbf{\Phi}}(\mathbf{x}(n))r^*(n)\}\end{aligned} \tag{F.11}$$

using (F.2), and since $e(n)\widehat{\mathbf{\Phi}}^*(\mathbf{x}(n)) = \widehat{\mathbf{\Phi}}(\mathbf{x}(n))e^*(n) - 2\Im\{\widehat{\mathbf{\Phi}}(\mathbf{x}(n))e^*(n)\}$ (based on properties from Sect. 5.1.2).

Comparing these, and since $\exp(-|e(n)|^2/2\sigma^2) < 1$, we see that the Quat-KMC convergence is slower than Quat-KLMS initially. When the algorithm approaches steady-state, and $e(n)$ is close to measurement noise $r(n)$, then $\exp(-|e(n)|^2/2\sigma^2) \approx 1$, and Quat-KMC, Quat-KLMS convergence rates are similar. This is seen in Chap. 11.

# References

1. I. Steinwart, D. Hush, C. Scovel, An explicit description of the reproducing kernel Hilbert spaces of Gaussian RBF kernels. IEEE Trans. Inf. Theory **52**(10) (2006)
2. W. Liu, J. Principe, S. Haykin, *Kernel Adaptive Filtering: A Comprehensive Introduction* (Wiley Publishers, 2010)
3. J.C. Principe, N.R. Euliano, W.C. Lefebvre, *Neural and Adaptive Systems: Fundamentals Through Simulations* (Wiley, 2000)
4. W.D. Parreira, J.C.M. Bermudez, C. Richard, J.-Y. Tourneret, Stochastic behavior analysis of the Gaussian kernel-least-mean-square algorithm. IEEE Trans. Signal Process. **60**(5), 2208–2222 (2012)
5. V.I. Paulsen, An Introduction to the Theory of Reproducing Kernel Hilbert Spaces. http://www.math.uh.edu/vern/rkhs.pdf
6. C. Cheong Took, D.P. Mandic, A quaternion widely linear adaptive filter. IEEE Trans. Signal Process. **58**(8), 4427–4431 August (2010)
7. S. Haykin, *Adaptive Signal Processing*, 4th edn. (Prentice-Hall, 2002)
8. P.S.R. Diniz, *Adaptive Filtering: Algorithms and Practical Implementation*, 3rd edn. (Springer Publishers, 2008)

## *Uncited References*

9. B. Chen, P. Zhu, S. Zhao, J.C. Principe, A sequential optimization framework to adapt the kernel size in the kernel least mean square algorithm. IEEE Trans. Neural Netw. Learn. Syst. (2012)
10. Quaternions Wikipedia, https://en.wikipedia.org/wiki/Quaternion
11. N. Wolchover, The peculiar math that could underlie the laws of nature, Quanta Magazine (2018), https://www.quantamagazine.org/the-octonion-math-that-could-underpin-physics-20180720



T. K. Paul and T. Ogunfunmi, *Adaptive Filter Algorithms For Complex and Quaternion Data*, https://doi.org/10.1007/978-3-032-23798-9

# Index




T. K. Paul and T. Ogunfunmi, *Adaptive Filter Algorithms For Complex and Quaternion Data*, https://doi.org/10.1007/978-3-032-23798-9

MIX
Papier aus verantwortungsvollen Quellen
Paper from responsible sources
FSC® C105338

If you have any concerns about our products,
you can contact us on
**ProductSafety@springernature.com**

In case Publisher is established outside the EU,
the EU authorized representative is:
**Springer Nature Customer Service Center GmbH**
**Europaplatz 3, 69115 Heidelberg, Germany**

Printed by Libri Plureos GmbH
in Hamburg, Germany